K

# ENVIRONMENT, SECURITY AND UN REFORM

D0721447

## DATE DUE

*Also by Mark F. Imber*

\*THE USA, ILO, UNESCO AND IAEA

*\*Also from St. Martin's Press*

# Environment, Security and UN Reform

**Mark F. Imber**
*Lecturer in International Relations*
*University of St Andrews*

St. Martin's Press          New York

First published in the United States of America in 1994

Printed in Great Britain

ISBN 0–312–12168–7 (cl.)
ISBN 0–312–12169–5 (pbk.)

Library of Congress Cataloging-in-Publication Data
Imber, Mark.
Environment, security, and UN reform / Mark F. Imber.
p.  cm.
Includes index.
ISBN 0–312–12168–7 (cloth). — ISBN 0–312–12169–5 (paper)
1. Environmental degradation—Developing countries.
2. Environmental responsibility—Developing countries.
3. Environmental policy—Developing countries.   4. United Nations
Environment Programme.   I. Title.
GE160.D44I43   1994
363.7'009172' 4—dc20                                          94–5974
                                                                          CIP

To A. J. R. Groom and John Simpson,
whose enthusiasm and excellent teaching most
encouraged me to study international organisations

# Contents

# List of Tables

# Preface

Why another book on environmental diplomacy? The issue of environmental security has been addressed very well in many places, by Arthur Westing, Caroline Thomas, Ian Rowlands, Daniel Deudney and others. The linkage between environmental degradation and the burdens of Third World debt has also been addressed, by Morris Miller, Susan George and others. Proposals for the protection of the global commons have been advocated since the 1980s by Willi Brandt and Gro Harlem Brundtland in their world-famous reports, and by Barry Buzan, Steven Krasner and John Vogler, in the academic literature. I am not, however, aware of a *single* study which attempts to synthesise this triple dilemma, within the analytical framework of international relations and furthermore, linked explicitly to the agenda of UN reform.

In 1989 I concluded a study of the troubled relations between the US and the specialised agencies with the observation that the environmental agenda would prove the toughest test of the UN system's ability to recover from a decade of alleged politicisation and the American-led response of budgetary austerity and partial boycott. This study, although standing alone as an investigation of environmental and security issues, represents, for the author, a second stage in the study of the UN system, linked by a concern, common to both studies, to investigate the limits of the possible in what David Mitrany called 'the progress of international government'. As a rationalist and agnostic in world of passionate fundamentalists, the functionalist is born to be disappointed. However, in the world of states, 'binding together those interest which are common, where they are common and to the extent that they are common' is, alas, the only way forward.

In a sense this book was written backwards: each chapter initially chosen as the point of introduction needed explaining by one that logically had to precede it. From a nuts-and-bolts investigation of the mandate and performance of the United Nations Environment Programme, UNEP, commissioned by the ESRC in August 1990, it emerged that UNEP could only be explained if prefaced by discussion of more fundamental questions of the context in which the post-Stockholm development of the international environmental regulation had been conducted.

In trying to understand the causes of diplomatic inertia, intransigence and the persistently egotistical behaviour of states caught in the two 'prisoner's dilemmas' of environmental security and military security, I was repeatedly impressed by the centrality of financial arrangements as an underlying factor. For Third World countries their burden of debt is critical to their inability to embrace sustainable development policies. For the North, fear of the loss of competitive advantage, the recessionary condition of industry since 1990 and, above all, a decade of Mrs Thatcher and President Reagan, led hostility to regulation and intervention in the market-place, have inhibited the recognition of enlightened self-interest in address-ing the need for higher environmental standards.

Much that is written here will be familiar to the NGO and development studies community, but it is not common knowledge within the inter-national relations community, generally concerned with diplomatic and military relations. Herein lies another justification for persevering with what I admit is a work of synthesis. I am concerned to present to an IR readership the argument that environmental questions and UN reform are central and not peripheral to the field, and that innumerable 'security' ques-tions cannot be understood without adopting an environmental perspective.

# Acknowledgements

I am very grateful to the ESRC for financial assistance received in 1990–92. Peter Usher, David Lazarus and Gerry O'Dell at UNEP were especially helpful with the initial project, also staff in the UK High Commission, Nairobi, and officials in the Department of Environment and Foreign and Commonwealth Office, with particular responsibility for UNEP. Some tables and explanation contained in Chapters 5 and 6 appeared first in articles by the author, viz. 'Too Many Cooks? The post-Rio reform of the UN', *International Affairs*, Vol. 69, no. 1, 1993, and 'The UN role in Sustainable Development', *Environmental Politics*, Vol. 2, no. 4, 1993, and are reproduced with permission. Academic credits can be invidious. The memory is weak particularly when chance conversations may send an author off on a trail without recalling the original stimulus. That said, my thanks are due, in approximately chronological order, to; A. J. Groom, Sally Morphet, Mohammed Malek, Anthony Clunies Ross, Ian Rowlands, Jacqueline Roddick, John Vogler, Peter Willetts, all the members of the BISA, GEC group, Michael Grubb, David Scrivener, Brian Clark, Eric Jensen, The Mono Lake Committee, Caroline Thomas, Roderick Ogley, Illiam Costain, and my colleagues at St Andrews, especially Trevor Salmon, for supporting my research leave in October–December 1992.

I am also grateful to a number of personal friends: Cathy Davies, Roy and Cecilia Dykhof, Richard Fardon, Guy Perry, Debbie and Steven Poole, Lorna Smith and Tricia Waite for their friendship and encouragement while I was writing this book. The plan of this book became clear one brilliant Sabbath afternoon in August, whilst taking the Caledonian MacBrayne ferry-boat from Uig to Lochmaddy. I'm not sure whom to thank for that. In an academic system of performance indicators and research selectivity exercises it's nice to know that you can still get a good idea leaning over the side of a boat watching the Minch slip by.

# 1 Two Hiroshimas Every Week

Then God said, 'Let us make man in our image, after our likeness; and let them have dominion over the fish of the sea, and over the birds of the air, and over the cattle, and over all the earth, and over every creeping thing that crawls upon the earth.

Genesis 1: 26 *Revised Standard Version*

I'm truly sorry Man's dominion
Has broken Nature's social union,
An' justifies that ill opinion,
Which makes thee startle,
At me, thy poor, earth-born companion,
An' fellow mortal!

Robert Burns, *To A Mouse*

Nuclear weapons have not been exploded in anger since August 1945. The bombing of Hiroshima and Nagasaki directly killed approximately 100 000 persons. Today, the effects of poverty – preventable diseases and hunger – kill 12.9 *million* Third World children every year. Poverty therefore kills Third World children at the rate of an Hiroshima and Nagasaki bombing every three *days*.[1] This figure can also be expressed as reproducing the six million Jewish dead of the Holocaust every six months. Put yet another way, the world at peace is suffering a rate of child-mortality that exceeds the death-rate of the Second World War. Poverty kills nearly 70 million children over five-and-a-half years. (Compare 58 million dead in the war.) Such avoidable casualties in the struggle for development focus the western mind to reconsider just what is understood by both the terms security *and* environmental quality, and how the two might be linked.

The most common cause of preventable child-death is diarrhoea. Simple infections cause massive dehydration which can be readily treated with simple salts and glucose mixed with *clean* water. In the developing countries one estimate suggests that 'more than 95% of urban sewage is discharged into surface waters without treatment'.[2] *Agenda 21*, the United Nations action plan adopted at Rio in June 1992, estimates that '80% of all diseases and over one third of all deaths in developing countries are caused by the consumption of contaminated water'.[3] In Britain, the public debate on water quality standards and the privatisation of utilities has shown that

1

although rain may fall freely from the sky, collecting, storing, cleaning and piping drinking-water is a multi-billion-pound industry. The same *Agenda 21* has estimated the cost of bringing potable water to all in the developing countries at $20 billion per annum between 1993 and 2000.[4]

In the period 1984–90 the flow of aid and investment from the developed countries of the North to the developing countries of the South has been more than reversed by the flow of interest and capital repayments by developing countries on their previously acquired debts. Through those seven years, the developing countries sustained the developed countries by this flow of income, during which time the popular headlines in the UK tended to emphasise the bankers' provisions for Third World default, and thus sought to shift the blame for interest-rate policies and the tough treatment of personal customers during the UK recession on to the imagined lenient treatment of the poorest countries in the world. The annual net transfer was $40 billion dollars in 1990. The total over the period 1984–90 has been estimated at $200 billion. In one author's estimate, this represented a transfer in real terms (i.e. allowing for exchange-rate fluctuations and inflation over the period), of *six* Marshall Plans, given *by* the developing countries to the rich countries of the North.[5] (The Marshall Plan transferred $16 billion to Europe in 1948–50.) At the bilateral level the aggregate of official development assistance (ODA) extended by the OECD countries in 1991 ran to $56.7 billion. Despite the 20-year-old commitment to raise ODA to the target of 0.7 per cent of GNP, the average for the OECD is almost exactly half of this target, i.e. 0.34 per cent. Implementing the 0.7 per cent target, as reiterated at Rio UNCED, in June 1992, would thereby effectively double the ODA to approximately $125 billion, and so just restore a positive flow of net transfer from North to South.[6] However, it is clear that ODA alone is not a sufficient answer to the linked problems of debt and environmental degradation, the failure of development and the intractability of many countries' security dilemmas.

Numerous territorial disputes, especially in arid regions, involve underlying tensions over water-rights. Israel, Jordan and Syria dispute access and drawing rights in the Yarmuk–Jordan system. Israel's reluctance to give up its occupation of south Lebanon is complicated by the presence of the Litani river. Israel's Palestinian dilemma is compounded by the future use of the West Bank aquifer. Turkey, Syria and Iraq are in dispute over the Tigris–Euphrates system or, more particularly, Turkey's new control of the headwaters of the rivers gained by the construction of the Ataturk dam. Turkey's internal ability to confront its Kurdish question is similarly affected by the dam which lies in the Kurdish provinces. Ethiopia, the source of the White Nile, is estimated to draw only 1 per cent of the river's flow for its own use. The civil war in Southern Sudan has been directly exacerbated by the attempted construction of the Jonglei canal, which aims to draw water from the swampland more quickly and with less evaporation, directly into the White Nile. These examples are typical of traditional, territorial and fixed-sum conflicts between states that derive in part from increased human demands and environmental degradation.

The intractability of such disputes, and the impact of a continued net transfer of capital from the developing countries to the industrialised countries, cited earlier at circa $40 billion per annum, quite dwarfs the efforts of UN agencies to create structures of international government. The most optimistic forecast of expansion in the United Nations Environment Programme (UNEP) bi-ennium budget is to $250 million; the UN Development Programme (UNDP) may rise to $1 billion; and the World Bank funded Global Environmental Facility (GEF), originally resourced at $1.3 billion, may rise to $2.5 billion. When the conditions attached to these figures are taken into consideration, the flow is even less impressive. For example, GEF money is only available for projects deemed to be of global importance, specifically for actions to address greenhouse gas emissions, the preservation of biological diversity, the protection of international waters and the ozone layer.[7] Despite the need for clean water, Third World city sewers are a strictly local matter.

The United Nations has, since 1990, been burdened with the expectation of delivering a post-Cold War, New World Order. This has been emboldened, but also narrowly interpreted in terms of *military* security, by the novelty of a Security Council consensus creating new peacekeeping operations and legitimating the coalition action in restoring Kuwait's sovereignty by military means in 1991. The Cold War hiatus has also allowed the international community to address questions of environmental degradation and the faltering progress of development; and especially the linkage between them. These expectations were centred on the United Nations Conference on Environment and Development (UNCED), convened at Rio in June 1992. However, the UN's capacity to act on both the military security and UNCED agendas has been crippled by the enormous financial shortfall between the members' rhetoric and their willingness to fund that rhetoric. The four years of enhanced demands saw the UN financial crisis, caused by late payments and outright default, deepen throughout 1990–93. During the summer of 1993 UN estimates of monies owed to it stood at a total of $1.7 billion.[8]

Child-mortality rates, the net transfer of capital from the poor to the rich, the declining military security of many Third World societies and the inadequacies of current United Nations structures constitute the islands of comprehension which this study seeks to connect together. Governments crippled by debt find it increasingly difficult to devote resources to maintaining and improving local environmental standards, let alone to upholding their international obligations. States which are confronted by irreversible environmental degradation in the form of soil erosion, deforestation, desertification, the parlous state of public health provision, urban blight and rates of population growth that will double their population in less than 30 years (let alone the more exotic possibilities of rising sea-levels and the medical effects of ozone-layer depletion), may seek to resolve their difficulties by the acquisition of neighbouring territories by force, or simply by the pressure of mass migration. The EC and the members of NATO, as Mediterranean littoral countries, will not be immune to these processes of change, some of which are at their most acute in the Middle East and North Africa. Yet, incredibly, *Agenda 21* assumes that the developing countries will fund the programmes adopted at Rio to the extent of $475 billion per annum from their *own* resources, to which will be added $125 billion that will flow in ODA from the OECD region.[9]

The reform of the international governance, that is the global environmental, financial and security apparatus which is necessary to reverse this vicious circle, relies upon the adoption of so-called sustainable development practices. 'Humanity has the ability to make development sustainable – to ensure that it meets the needs of the present without compromising the ability of future generations to meet their own needs.'[10] Briefly summarised, the environmental and developmental malaise of the Third World can be addressed by measures which link the several issues identified here. Massive concessions by creditor nations on the debt issue, combined with the imaginative use of debt-for-nature conversions of the remaining debt, are necessary for most indebted countries. The UN system would benefit from measures to raise *international* revenues by taxing the use of the global commons (the atmosphere, its climate system, the high seas and the sea-bed beyond the limits of national jurisdiction), and by taxing *transboundary* environmental pollution. Funds need to be generated by more imaginative means than simply shaking the tree on which the 0.7 per cent ODA target should bear fruit. The burden of both domestic taxation and international charges could be *shifted* from subsidising environmentally destructive and non-sustainable practices, such as nitrogen-intensive farming which produces surplus commodities, and towards environmentally protective, restorative and sustainable practices. The elements of such a scheme might include carbon taxes on primary energy use, offset by tax-credits for forestry management, i.e. a comprehensive 'carbon-banking' policy relating to sources and sinks of carbon dioxide. The UK's mooted introduction of 17.5 per cent VAT on domestic gas, electricity and fuel oil, has been justified on environmental criteria. Such claims would make more sense if VAT were simultaneously *removed* from or lowered on sales of thermal insulation materials, secondary glazing, heat-recovery systems and other energy-efficient products.

The expansion of ODA, also the creation of new financial resources derived from charging the full economic rent of the traditionally 'free' commons, can be combined with substantial reform of the UN system, itself nearly fifty years old. In this way Third World countries might be assisted to resolve their own local and regional environmental crises. However, sustainable development is not only for the South. The widespread adoption of environmentally sustainable policies in the North is required to slow the rates of consumption of non-renewable resources and to conserve the quality of the global commons. Despite the population imbalance between North and South, when measured over the last two hundred years of industrialisation and colonialism, it is the rich countries that have impacted most upon the capacity of the commons to sustain life on earth. Easing the debt-environment linkages will also provide a helpful path to ease the many military-security anxieties that these countries face. Many new environmental problems are symptoms of old resource disputes. The struggle between states for oil, soil and water is not new.

## ENVIRONMENT AND THEORIES OF INTERNATIONAL RELATIONS

What insights does international relations theory offer which might explain how states address the issues raised above? More problematic is to explain the apparent *reluctance* of the scholarly and diplomatic communities to incorporate the environmental agenda into the mainstream of international relations in the same way that the military-security agenda and *selective* aspects of political economy, most obviously trade and financial relations *between* the developed countries, are clearly regarded as central to the discipline. One explanation concerns disciplinary boundaries. Most of the issues discussed in this study are more typically explored in the literature of development studies. So long as the international relations community defines itself in terms of studying primarily military-security questions, the 'eyes' of the discipline simply will not 'see' environmental causes of profound instability in the international system. How has this situation arisen? Each of the mainstream approaches to international relations (realism, pluralism and structuralism) is in truth only 'partially-sighted' with respect to environmental issues.

Realism is primarily concerned with the protection of state-sovereignty and the survival of the state as a discrete actor. This perspective tends to rank environmental questions comparatively low on the scale of threats to national security, relative to the capabilities and intentions of potentially hostile *human* enemies in neighbouring states. If an environmental problem is recognised as a potential threat to the security of the state, the territorial and possessive qualities of the realists tend to favour unilateral solutions to such problems, seeking to maximise, or at least optimise the state's access to scarce resources of water, oil and soil. Realism advocates action by the state and its agencies to maintain and/or acquire control of resources. Faced with threats to the continued enjoyment of such resources, realism endorses military actions to maintain the state's advantage. Examples of such tough strategies would include numerous cases of military intervention to maintain control of strategic resources.[11] The preference of the Reagan and Bush administrations for the privatisation of the global commons, specifically mining-rights on the seabed, would conform to the realists' desire to maximise free access to the commons by those with the financial and technical resources to exploit them: 'first-come-first-served'; in preference to a common-heritage regime with commitments to redistribute income among those at 'the back of the queue'. The nationalist rather than internationalist imperative associated with realism was also seen in the preference of the Bush administration to ride out the effects of climate change, reflecting conservative estimates of its impact on American GDP, rather than funding drastic preventive measures.[12]

Basing its overall world-view on identifying complex patterns of interdependence in the world economy, pluralism embraces international organisation and international law as the preferred instruments for managing disputes between states. This includes environmental issues. For the pluralist the environment is an issue-area akin to monetary or trade questions. Optimal solutions for parties that negotiate in good faith can be secured by the appropriate mixture of formal and informal rules that constitute a regime.[13] Pluralism argues the need for egotistical sovereignties to manage their competitive instincts within a framework of rules, if not actual cooperation. The approach is receptive to the idea that transnational and genuinely global threats to the stability of the state system may exist, which cannot be defined in military terms. The pluralist would point to the large number of comprehensive and binding multilateral treaties between states which already serve to maintain environmental quality, as evidence of the ability of the international system to respond to this new challenge. The Trail smelter arbitration of 1939, which established the liability of states for the effects of cross-boundary pollution, was one such agreement.

Others would include the North West Atlantic Fisheries agreement, the many conservation agreements attaching to the Antarctic Treaty and the recent flurry of environmental diplomacy on ozone-layer protection, biological diversity and climate change.

The structuralist approach to international relations regards environmental degradation as an inevitable characteristic of the capitalist worldsystem. This approach argues that four hundred years of imperialism, colonialism and now neo-colonialism have created an unequal world in which poverty, racism and environmental degradation are reinforced by market forces. These serve to depress commodity prices and so perpetuate a cycle of undervaluing the natural wealth of the developing countries.[14] Furthermore, transnational corporations will seek the most permissive regulatory regime for their operations, and so expose the weak and the poorlyorganised to the impact of polluting industrial processes. Thereafter, military–political intervention and indebtedness will together perpetuate the dependency of the developing countries. The legal and institutional mechanisms which pluralists favour to encourage environmental conservation are regarded by structuralists as inadequate mechanisms for change or, worse, as vehicles for co-option and continued exploitation within the increasingly unified global economy, or so-called world-system. The world-system suggests a global division of labour in which the traditional metropolitan loyalties of corporations to some notion of a home territory and workforce is replaced by a cosmopolitan world-view, which locates investment, manufacturing and employment wherever comparative advantage dictates. This contains a particular threat to the environmental quality of the Third World, as regulatory zeal and rising skilled labour costs in Europe, North America and Japan increasingly tempt multinational enterprises to locate dirty, polluting, dangerous and low-wage stages of production in poor countries which may even welcome them. Toxic-wastedumping in West Africa, the Bophal disaster and the state of the Mexican *maquilladora* would attest to this interpretation. Despite a powerful diagnostic insight, structuralism is less plausible in its prescription, advocating the revolutionary transformation of the world economy.

Each paradigm is limited in its explanatory powers. Realists tend to have a legalistic, almost simplistic view of sovereign equality. They hold the governments of newly-independent countries responsible for whatever obligations, debts and problems they may inherit and encounter, as if the legacy of the past four hundred years could be eliminated by the ceremony of lowering one old flag at midnight, and raising another of gaudier stripe in its place. Structuralists are prone to underestimate, or to denounce on grounds of political correctness, the suggestion that persistent environmental degradation and Third World poverty may in part be attributed to the effects of fatalistic cultures and religions, or to government corruption and high rates of population increase. Pluralism tends to play down both the centrality of national interests and the historical asymmetry of rich and poor states as obstacles to negotiating binding regimes on environmental questions. Pluralists therefore overestimate the ability of states, as rational actors, to use international organisations and international law to make sacrifices for the greater good. It is now appropriate to explore the limitations of the three approaches to international relations more thoroughly.

## REALISM AND ENVIRONMENTAL SECURITY

There is clearly an environmental dimension to military security. If no more than a rewording of the literature concerning access to natural resources, the acquisition and defence of strategic resources have always been central to concepts of military security. If the extended agenda of global environmental change is incorporated into the security dilemma of states, then non-sustainable patterns of economic development and exchange require sustainable solutions. Contrary to the spirit of the Brundtland report, realists would assert policies in defence of the national interest, which might involve military options. The environmental dimension may therefore be defined as some new *genus* of security, or more conventionally to the realist, environmental threats may be subsumed within an enlarged, conventional definition of military security. This debate, involving what has been called the 'encompassing' or 'add-on' approach, is well discussed by Caroline Thomas.[15]

The case for redefining security comprehensively is reminiscent of John Herz's 1959 hypothesis on the decline of the territorial state. In Herz's case this was caused by the advent of intercontinental ballistic missiles, which, before SDI, were unstoppable. Their very existence undermined the Hobbesian justification of the state, as an agency able to extend its protection against external aggression.[16] The decline of the territorial state has been a shibboleth of many subsequent analyses of international relations, including the advocates of transnationalism and other varieties of integration theory. In many ways the threats identified by exponents of the environmental threat to national security exhibit the same characteristics. Diverse environmental threats to the security of the state are pervasive, irreversible and, by their transboundary nature, undermine the concept of territorial defence; e.g. the medical effects of ozone-layer depletion, property damage by acid-deposition, shifts in agricultural productivity induced by climatic change. Among the more obvious environmental threats to conventional, territorial concepts of security are the most modest of predictions of sea-level rise (about 0.5 metre by 2050), and continuing desertification attributable to global warming. Sea-inundations, and rising salt-water penetration of the water-table will render agriculturally crucial areas such as the Nile Delta and the coast of Bangladesh less habitable.[17] Sea-level rises may trigger a second problem, the mass migration of environmental refugees, akin to the 1970–71 movements from the East Bengal cyclone disaster that became one of the contributory factors in the Pakistan civil war, Indian intervention and the creation of Bangladesh. Egypt, Bangladesh and the Pacific and Caribbean island micro-states, and hence their neighbours, may witness unprecedented refugee migrations. The movement of millions of sub-Saharan Africans arising from the desertification of the Sahel during the last decade may be eclipsed by the magnitude of these new categories of refugees. The current concept of refugee status is essentially political and based upon 'a well-founded fear of persecution'. This is inadequate, both conceptually and legally, to grasp the issue. Mass migration of unarmed *environmental* refugees may be regarded, bureaucratically, as an immigration problem rather than an invasion; it is less likely that citizens in the receiving states will note the distinction.[18]

Competition for water resources in shared river systems, such as the Jordan, Yarmuk, Colorado, Indus and Euphrates, creates tension between states, especially those with escalating water demands due to both population growth and industrial modernisation. The water resources of disputed territories, such as groundwater pumping in the West Bank, are crucial to any long-term territorial settlement in the Middle East. Other environmental challenges to territorial security include the need to police and enforce fishery conservation in the 200-mile Exclusive Economic Zones, a particular problem for countries without the necessary naval power to uphold their claims. The subversion of small island-state governments by larger regional powers seeking access to fishing grounds and safe-basing facilities is also possible. In the late 1980s, Tuvalu was seriously courted by the USSR for this purpose. (For reasons that realists have yet to explain, the unarmed micro-state has survived and the nuclear-armed superpower with a blue-water navy dissolved itself, peacefully, on 31 December 1991.) Finally, targeting environmentally sensitive facilities and industrial processes might feature in both conventional military doctrine and as terrorist targets.[19]

The corollary of these linkages, which all emphasise environmental threats to the military security of the state, is the long-established linkage that certain military activities are a threat to the environment. Arthur Westing has documented the use of military tactics and strategies that have deliberately targeted, or inflicted collateral damage on the environment. Westing discusses herbicide damage, nuclear testing, and possible nuclear weapons use, its blast and fall-out effects, the nuclear winter hypothesis, weather modification and deliberate flooding either offensively, as in the British attacks on the Ruhr dams, or for defensive purposes as in the Chinese and Dutch destruction of their own flood defences in 1938 and 1940 respectively.[20] The deliberate destruction of oil facilities in Kuwait, by Iraq, prior to their retreat in April 1991 introduced a new concept of environmental warfare. It not only sought to cause damage to the enemy, but perhaps also sought to deter the coalition's response, in a sort of environmental blackmail, in which TV coverage of oil-soaked birds played on the emotions of the watching public. The most alarming predictions of monsoon disturbance caused by the burning of the wells, and of marine devastation in the upper Gulf caused by the slicks were thankfully not substantiated. The local and short-term impacts were severe, but as one summary of UNEP's own 50-member survey team reported, 'from those areas accessible to scientific scrutiny, only a few have experienced irreversible disruption as a result of the war'.[21]

The state's provisions for military security do not only impact upon the environment in the direct ways discussed above. Military *expenditure*, or more properly the alternative uses to which it may be put, raise major issues of so-called military conversion. The field of military conversion is complex and tendentious, but the sums involved are so large that even modest economies in the provision of military expenditures can release significant resources for the provision of environmental security. The Brundtland report was published at the height of post-Afghanistan tensions between the superpowers and cited a figure of $900 billion global defence expenditures for 1985. Implementing the recommendations of the UN Water and Sanitation Decade programme was costed at $30 billion per annum. The provision of contraceptive materials to all women wishing to use them was costed at $3 billion.[22] These are examples of how the definition of security can change under the impact of greater environmental awareness. They also illustrate the entrenched nature of the military-security mind-set, in that 'military-conversion' remains in the ghetto of the political left rather than being accepted as part of an encompassing redefinition of security.

Deudney in fact argues *against* linking the language and methods of threats to national security and environmental degradation. He maintains that the nature of the threat, the enemy, their intentions and the timescales involved, all distinguish threats presented by environmental degradation from the normal discourse of national security. He therefore warns against adopting the language of the moral equivalent of war.

Table 1.1    Attitudes to security

| National Security | Global Habitability |
| --- | --- |
| specific threats | diffuse threats |
| enemy as 'others' | enemy as 'ourselves' |
| intended harm | unintended harm |
| short time-scales | long time-scales |
| mainly zero-sum | common benefits |

*Source:* adapted from D. Deudney, 'The Case Against Linking Environmental Degradation and National Security', *Millennium*, 19, 3 (1990), pp. 464, 466.

Deudney argues that the robustness of the modern world trade system, the substitutability of modern materials science and the intractability of occupying territory in the face of guerilla resistance, renders unnecessary military aggression by the developed countries to secure environmental resources. Duedney also downgrades the likelihood of the global under-class rising as a military threat to the North, driven by their search for habitable space after desertification, agricultural collapse or population pressures overtake them. Quoting Bernard Brodie, he observes that 'the predisposing factors to military aggression are full bellies not empty ones'.[23] Maintaining a minimal nuclear deterrent would insure against such an eventuality. Deudney is perhaps too sanguine in this argument. While it is perhaps unlikely that the global poor will respond to their plight with armed aggression, the threat, if threat it be, is more likely to be manifest as mass migration. This movement is more likely to be measured over several decades, and will more properly test the *internal* security agencies, such as immigration control, police, humanitarian and refugee relief agencies, more than the guardians of *external* security.

Thomas, consistently with her other writings, is most concerned to establish the linkage between environmental degradation and the internal legitimacy and stability of state authorities in the Third World. Using Jackson's distinction between juridical and empirical statehood, she emphasises the vulnerability of many governments.[24] More typical of the 'encompassing' argument is that presented by Evteev, Perelet and Voronin, who expressly link environmental diplomacy to the question of security. It is an initially familiar review of the problems requiring international attention, but concludes with an increasingly specific set of principles for 'ecological security'. The authors include the conservation of the commons, environmental impact assessment for development projects, appropriate technology, global data bases and information exchange, arrangements for the notification of transboundary pollution risks and incidents, and commitments to settle transboundary environmental problems by peaceful means.[25]

This discussion has demonstrated that on a number of issues, such as the defence of natural resources, the realist preference for a conventional sovereign-territorial approach to environmental threats is still valid. However, no amount of unilateral action within the realist framework will protect the stratospheric ozone layer. No appeal to patriotism can keep one country's sea-level constant, while for others, of less zeal, it rises. Even if a particular source of harmful trans-boundary pollution can be identified, it is unlikely that states will sanction military force to demolish the offending chimney. If the source of pollution is a leaking nuclear reactor, like Chernobyl, such action would be self-defeating. A growing number of environmental threats to the security of the state cannot be solved by unilateral action, but *can* only be approached through multilateral bargaining.

## THE ROLE OF INTERNATIONAL ORGANISATIONS

The literature on the role of international organisations in environmental issues, written between the early 1970s and 1990, is not extensive. Kay and Skolnikoff (1971), Boardman (1981) and Kay and Jacobson (1983) were exceptions.[26] In the 1980s, a series of UN and US reports such as Brandt, *The Global 2000*, and the Brundtland Report, all sought to explore the linkages between economic development and environmental quality. Each envisaged an expanded competence for international organisations both as forums for the negotiation of agreements and as executive agencies for administration in the functionalist manner.[27]

The pluralist approach to multilateral diplomacy claims that conference diplomacy is the only arena in which governments can make agreements which are:

- multilateral (involving three or more states),
- public (adopted by vote in plenary sessions),

● adopted simultaneously (to overcome the inhibitions of unilateral actions), and which are
● preferably, but not necessarily, legally binding.

A public expectation of rapid international agreement on the complex agenda of environmental questions was raised by the scientific consensus and political timeliness displayed in negotiating the Vienna Convention and Montreal Protocol on substances harmful to the ozone layer. However, for several reasons discussed later, the CFC question has since proven to be the exceptional case. Negotiating other agreements has involved greater conflicts of scientific judgement and of national interests. These conflicts have slowed the process of negotiation and reduced the value of the agreements made. How can these obstacles be explained?

The absence of clear priorities in the environmental agenda, the existence of multiple channels of communication between parties on these issues, involving not only governments but also IGOs, and INGOs, and the near-irrelevance of military force to secure a solution to these issues, all illustrate the conditions of so-called complex interdependence, identified by Keohane and Nye.[28] The complexity and interconnectedness of many environmental problems creates paralysis in policymaking. The absence of any clear hierarchy of goals and priorities is typical of the sectoral nature of many environmental problems, and inhibits comprehensive agreements. For example, there is a destructive cycle of environmental pressures linking Latin American patterns of land-tenure, the rate of tropical deforestation, rising carbon dioxide emissions, the loss of biological diversity and climate change. Each, it can be argued, is the appropriate issue on which to *initiate* negotiated change towards sustainable practices. However, when numerous interests are involved, no one party has any obvious incentive to make unilateral concessions. Economists and game-theorists would recognise the familiar problems of the 'free-rider' and the 'prisoner's dilemma'.

The free-rider describes anyone who benefits from public services without paying the fare or tax for which they are liable. One free-rider does not bankrupt the bus company. However, widespread fare-dodging or tax-evasion does begin to undermine the revenues needed to maintain that service. As rational actors, many citizens try to take a 'free ride', and then complain that the quality of public services consistently falls short of expectations and demand. ('The man who does *not* cut wood in a state forest is a fool' – contemporary Turkish proverb.) Public goods are typically those goods and services which cannot practically collect a user-fee at the point of consumption. Street-lights which required a coin-in-the-slot as drivers went past would be inconvenient. Once they have been provided, public goods provide benefits which are indivisible. Street lights do not know to turn off when a council-tax-dodger or joyrider drives past. For these reasons, public goods are usually provided by a public authority, which can enforce payment through taxation and reap cost-savings through the large-scale provision of what, before 1979, were unblushingly called 'natural monopolies'.

Just as street lights do not switch off when the tax-dodger passes by, so lighthouses serve the merchant-ship and drug-trafficker equally well. Some traditional neutrals derive security from the alliances which they shun; many countries benefit from UN peacekeeping activities to which they do not contribute either manpower or cash. The free-rider is in many ways a more persistent problem of international society than of domestic society. In the *international* anarchy of sovereign states, the absence of government has traditionally explained the poor provision of the appropriate *international* public utilities. This gap was partly filled, after 1815, by international organisations, originally created to manage large projects of international infrastructure such as the International Telegraph Union (ITU), and the Universal Postal Union (UPU). The trend was continued after 1945 by the expansion of the UN system of specialised agencies in fields of technical cooperation, economic development and humanitarian relief. International organisations can restrict membership to those countries which pay their annual subscriptions; they can vote to apply sanctions to countries that abuse the service the organisation provides. They can, in short, create islands of international government in the otherwise anarchic sea of international relations.[29]

International action on environmental issues is also inhibited by the problem of *externalities*. Any cost that can be passed on by the polluter constitutes an externality. By definition, transboundary pollution is someone else's problem, although subject to increasing international regulation since 1936. The easiest pollution to pass on is that which pollutes the global commons, beyond any state's jurisdiction. For this reason the high seas, the atmosphere, the ozone layer, Antarctica and outer space are all vulnerable to abuse. The so-called 'tragedy of the commons' will be discussed in a later chapter.

Game theory, especially the 'prisoner's dilemma', yields the same inference. Egotistical actors, such as individuals, corporations and states, can behave rationally, in pursuit of their own interest, and yet create results which are to their own detriment, and to the detriment of everyone else as well. Equally rational egotists behave in the same way, e.g. one individual littering, driving on a congested motorway or panic-buying; one country releasing CFCs, raising carbon dioxide emissions, or fishing beyond the limits of the maximum sustainable catch, does *not* get away with it, because all egotists behave in the same way. We trip over litter, sit in traffic jams and run out of sugar, because millions of our fellow citizens behave as selfishly as ourselves. As nations we damage the ozone layer, enhance the greenhouse effect and exhaust the North Sea fisheries because we *think* we can each get away with selfish behaviour which we trust our fellows not to emulate. But they do.[30]

International organisations of the UN specialised agency variety, i.e. those of nearly universal membership, providing technical and economic services, attempt to charge the free-rider, identify and internalise the external costs and try to reign in the impulse of sovereign states to create prisoner's dilemmas. One way to advance these objectives which is implied by the public, simultaneous and binding procedures identified above, is for states to satisfy diverse interests by the use of package deals. These imply progress by consensus rather than by majority vote. If any one state *can* wreck an agreement, all must be brought on board. The agenda of the UN Conference on Environment and Development, discussed in Chapter 5, illustrated these characteristics. Different interests, different priorities and different costs encouraged states to seek the accommodation of a complex agenda through large package deals. The process is not new. It is familiar in the context of the twice-yearly European Councils and, perhaps ominously, it also occurred in the fourteen-year process necessary to negotiate the UN Convention on the Law of the Sea.

The progress of environmental diplomacy through multilateral channels can be observed in five stages. These represent the extent to which governments are willing to share sovereignty with the international organisation in question.

- Centralising information,
- agreeing norms,
- negotiating rules,
- enforcing rules,
- providing programmes.[31]

The evidence of progress through these stages is mixed on environmental questions. The Montreal Protocol on substances harmful to the ozone layer, negotiated in an unprecedented time-scale of three years between 1985 and 1988, contains far reaching proposals for enforcement and substantial compensation for developing countries adopting the more expensive technologies. In other words, the ozone issue passed through Jacobson's five stages in less than five years. However, climate change and biological diversity only reached the stage of loosely-defined framework conventions at Rio in 1992. Forestry proved so controversial that only the most anodyne and contradictory statement of principles could be agreed, not completing even the second of Jacobson's stages.

The potential to create new international organisations, to enlarge the competence and capacity of existing agencies, and the need for their improved coordination, all illustrate the relevance of the pluralist approach to environmental questions. However, the progress of multilateral diplomacy in the environmental field has also been frustrated during the last decade by a cluster of distractions associated with the alleged politicisation of the UN agencies. Politicisation, or more particularly the US response to it, explains much of the estrangement of the US from the UN system of agencies during the past decade. In brief, the ideological preference of the US for market-led approaches to UN system issues was compounded during the Reagan years by a range of disputes in the UN system associated with membership and credentials disputes, the defence of Israel, the pursuit of so-called extraneous issues through the agendas of the agencies, the imposition of zero-budgetary growth and disputes concerning staffing, funding and voting systems.[32] As part of the post-Cold War realignment, most of those issues and the behaviour associated with them lapsed, so creating a great opportunity for the UN members to use the system for negotiation on questions of global environmental change. But, as shown above, it is also in the nature of autonomous actors that they are inhibited from acting in the collective interest by a variety of impulses to selfishness. This remains the central limitation to the pluralist's progress, which must rely on the bargaining processes of international organisations to reveal the path of enlightened self-interest.

THE STRUCTURALIST CRITIQUE

Structuralism is a variety of critical theory in international relations. It is neo-Marxist in orientation and argues that persistent global inequality is the most significant structure in the international political system, which itself determines the behaviour of the actors within it. The pattern of contemporary international relations has been shaped by four centuries of imperialism. It is profoundly unequal in the distribution of wealth and income, and, despite their nominal political independence, the economies of the developing countries and the political loyalties of their governing elites are enmeshed in a capitalist world-system which continues to uphold a condition of *dependency.* Structuralism therefore provides a powerful denunciation of capitalism as a system necessarily based upon exploitation.

Exploitation not only damages the lives of the global poor, but also their natural environment. Natural resources and habitats are therefore com-modified and expendable in the interests of the developed countries and the multinational enterprises that have increasingly replaced colonialism as the vehicle of a continuing imperialism.[33]

The early writings on dependency were developed within the UN Econ-omic Commission for Latin America (ECLA), especially associated with Raul Prebisch. In seeking to explain the relative lag in Latin American economic growth, Prebisch argued that the terms of trade had been moving adversely against the region ever since the mid-nineteenth century. That is to say, the value of Latin American exports declined, relatively, against the cost of imports. The continent was thus trapped, exporting agricultural raw materials in a declining market and able to purchase fewer and fewer manufactured goods in exchange. The *dependencia* writers argued that this cycle of dependency was exacerbated in the Latin American case by the culture, class-structure and distribution of income, in which a cosmopolitan elite favoured the import of luxury consumer goods over capital goods, so further postponing industrialisation.[34] In this scheme, land reform and import-substitution became the recommended paths to growth during the 1960s.

Immanuel Wallerstein developed the 'dependency' model into a global explanation of the world capitalist system, in which states occupy positions in the core, the periphery or the semi-periphery of that system. This analy-sis reflects the globalisation of the world economy that has occurred since 1945. Although set free from colonialism and the protected markets that it offered, the vertical integration of economic activity became very advanced in a *global* system. The division of labour, freedom of movement of capital and corporate decision-making therefore take place in a seam-less, transboundary world economy. However, the ownership and control of the corporations, and hence the repatriation of profits, relentlessly ensure a net transfer of revenues to the developed countries of America, Europe and Japan. Confronted with the difficulty of explaining the spectacular industrialisation of the East Asian economies, Wallerstein identified an important role of semi-peripheral countries, typically the newly-industrialised countries (NICs), such as Taiwan, Hong Kong and Korea, which function as both exploiter and exploited in this scheme.

An important connection between structuralism and the pluralist scheme concerns the argument for 'hegemonic stability', advanced by Keohane. Regimes which the pluralist-functionalist analysis would argue are based on some notion of enlightened self-interest (regimes such as the IMF–IBRD-administered, dollar-financed international liquidity of the period

1944–71, or NATO, 1949–89), in fact rest upon the *hegemonic* interests of the great power which is principally responsible for supplying these services: over the last century, namely the United Kingdom and latterly the United States.[35] Empires and superpowers, have, by definition, extended reach. A superpower is one which has global interests, *and* sufficient, available military power to defend them. The provision of an infrastructure, necessary for their own rule creates semi-public goods which are available, at less than marginal cost, to other states. The international liquidity of the Bretton Woods years, and the *pax Americana* are obvious cases. Nineteenth-century British naval supremacy protected every other country's free trade, General Wade's military roads in the Highlands were brilliantly exploited by the Jacobites in 1745.

The structuralist critique is therefore comprehensive, and in some senses unanswerable; it is an entirely alternative *paradigm*. One particular aspect of global inequality and its impact upon environmental quality will form the basis of extended discussion in Chapter 2. This concerns the pervasiveness of Third World debt, and will explain how, since 1985, the net transfer of capital from South to North has indeed reversed the flow of ODA from North to South. Ironically, the debt issue vindicates a central theme of the structuralist thesis just at the historical moment that the explanatory power of *any* neo-Marxist scheme seems to be at its weakest in the West.


CRISIS? WHAT CRISIS?


Concern with environmental issues frequently emphasises their alleged characteristics of novelty and crisis. However, the environment is not a new issue in diplomacy, so much as an issue which has been treated narrowly on a sectoral basis, as single issues and fashion have demanded attention. It would be more accurate to suggest that treating the environment as an *integrated* concept has only become an established diplomatic procedure since the path-breaking Stockholm conference of 1972. This approach rose higher on the intergovernmental agenda in the late 1980s, most obviously in a somewhat frenzied and late response to the preparatory demands of the UNCED at Rio in June 1992. McCormick has shown how the so-called environmental movement, as a broad range of NGOs, has frequently *led* the process of directing governmental attention to focus upon controversial environmental issues such as whaling, ocean pollution, nuclear waste, nuclear-testing and the fate of the rainforests. Most of the UN specialised agencies have their origins in NGO campaigns of a previous era. Williams has demonstrated how nineteenth-century governments reacted slowly and selfishly to issues ripe for regulation such as public health, and migration, and were dragged reluctantly to the negotiating table.[36]

Characterising environmental issues as *crises* is also simplistic and misleading. Crisis is an appropriate description for some natural disasters and for some sudden and unforeseen disasters involving human culpability, such as the events at Chernobyl in 1986 and the Kuwait oil fires of 1991.[37] However, one well-established definition of crisis, and its implications for decision-making, requires three specific factors to be present; one concerns the level of threat, two factors concern the question of timing. Hermann suggests that a true crisis combines:

- a high level of threat,
- a short period of warning, *and*
- the need for a rapid response.[38]

If any one of these factors is absent or ameliorated this creates a variety of different situations, but not a crisis. This may appear to be academic hairsplitting, but several useful insights follow from adopting Hermann's scheme. Two of particular interest to environmental diplomacy are the situation in which a high level of threat is combined with the need for a rapid response, but in a situation which can be *anticipated* i.e. a long warning-time is possible. Hermann calls this a 'reflexive situation'. This assumes that the appropriate public authority has on-the shelf plans for an appropriate response, such as emergency planning procedures for evacuation, for rescue and disaster relief in areas of *regular* natural catastrophes. In practice, these may not be available because of several failures of the political process. There may be suppressed evidence of planning errors or technical malfunctions in man-made systems, or a political reluctance to confront issues of public safety. These situations are what campaigning environmentalists would describe as an 'accident waiting to happen'. With perfect hindsight, and the discovery of human failure we can all join in the denunciation. In this scheme Chernobyl and Three Mile Island, Seveso, Basel and Bhopal only became so-called 'crises' quite unnecessarily.

Many environmental issues present society with a well-anticipated, high-level threat, but allow an extended time for decision on the appropriate response. An example would be the climate-change question. Adopting the appropriate *preventive* responses may quite reasonably extend over several years, as in negotiating a climate-change convention to limit emissions of greenhouse gases. Meanwhile, *adaptive* responses, such as constructing sea-wall defences, the resettlement of endangered populations and crop research to develop salt-tolerant species, will be spread over several decades. Hermann calls this a 'deliberative situation'. In theory, an extended time for response can be used to *avoid* crisis, so long as remedial action is initiated well in advance of its *apparent* need. This is, of course, a particular form of the *precautionary principle*. As shown in discussing the limitations of conference diplomacy, it lies at the centre of each state's 'prisoner's dilemma'. To take unilateral, expensive actions in the field of environmental regulation may incur a loss of comparative advantage in trade and economic growth. (On the other hand, it may not. Japan has gained enormous competitive advantages over the United States by deliberately adopting energy-efficient industrial practices.) Furthermore, such actions may prove unnecessary if the threat recedes, or futile if the threat materialises, and for want of *collective* action the worst does indeed come to pass. The pay-off matrix may be shown as in Table 1.2.

Table 1.2  The environmental wager

|  | *Pessimists correct* | *Optimists correct* |
|---|---|---|
| *Adopt preventive action* | Modest sufficiency | Lower growth |
| *Delay preventive action* | Disaster | Higher growth |

Like Pascal's wager on the existence of God, the only temperate course, that which avoids disaster (that is, Hell in Pascal's version), is to adopt the precautionary principle (or lead the Godly life), in the full knowledge that this may incur a penalty in economic growth (pleasure and ease). This was characteristic of many of the UNCED agenda items. They were, and remain, genuinely global, genuinely threatening, but low on the political agenda for reasons of the extended time available for action. The *electoral* time-horizon, in many developed countries, inhibits action on any question that can be postponed for five years. In many developing countries, which may discount environmental quality at different rates, the short-term issues, such as this year's harvest, and the next monsoon's flood, tend to dominate the political and popular agenda. As the tragic condition of Bangladesh demonstrates, flooding, cyclone damage and cholera epidemics recur, and require actions over maybe fifty years to adopt sustainable practices in reforestation, flood-control and achieving the demographic transition to a birth rate that ensures replacement without growth. The precautionary principle must be applied internationally because the reforestation practices that would most assist Bangladesh are needed 500 miles to the north-west, in the Nepalese valleys, through which flow tributaries of the River Ganges. Here, population pressure, firewood gathering and overgrazing have combined to strip the forest cover and accelerate soil erosion, which creates flash-flooding and siltation in downstream Bangladesh.

Perhaps only the Greens and aristocrats operate on the necessary time-horizon. Greens and dukes plant oak trees for their grandchildren. The inter-generational factor is really the greatest 'externality' in costing the environment. The lack of inter-generational foresight is further complicated if action is needed in advance of full scientific consensus. The *uncertainty principle* undermines confidence. Discussing the North Sea, Saetevik has shown that political inhibitions exploit scientific uncertainty, contrasting,

on the one hand, the view that 'as long as no direct harmful consequences have been proved, it is desirable to use the sea as [a] waste recipient' and on the other hand, the view that 'until it has been proved that pollution has no harmful effects, this must be reduced in order to avoid possible irreversible damage to the eco-system'.[39]

Pascal's wager has brought many agnostics to a grudging faith in God, but the similar dynamics needed to achieve precautionary, regulatory decisions are not so obvious. Indeed, the inertia in the Bush White House in the years 1990–92 specifically cited the costs of premature action on the climate question.

Consensus is needed at both the technical-scientific level, and at the political level. This produces four possible situations, only one of which is likely to lead to the negotiation of binding environmental commitments.

Table 1.3 Achieving environmental consensus

|  | **Ozone** | *Issues*<br>*Forestry* | *HIV* | *Climate* |
|---|---|---|---|---|
| *Tech consensus?* | Y | Y | N | N |
| *Diplomatic consensus?* | Y | N | Y | N |
| *Outcome for regime* | Y | N | N | N |

In the case of action to protect the ozone layer, it took just 11 years to work from the Molina and Rowland article in *Nature* in 1974, which first discussed the possibility of CFC damage, to the Vienna Convention of 1985. This represents the sort of progress that is possible when scientific and political opinion converge in the desire for regulatory action. This was further demonstrated in the rapidly negotiated tightening of the original Convention by the subsequent adoption of the Montreal Protocol and London amendments. The willingness to fund HIV research has preceded any scientific consensus or breakthrough in the search for a vaccine. The known ecological damage done to the tropical forests is discounted in some countries and agreement frustrated, as shown at Rio in June 1992. Despite the consensus that emerged at IPCC during August–September 1990, it is clear that the US administration preferred, during 1990–92 to listen to the minority of climatologists dissenting from the IPCC position. These were assisted by Nordhaus's economic arguments, that the cost of riding out the worst effects of climate change might not exceed 0.26 per cent of GNP on an annual basis, while the cost–benefit ratio of drastic intervention to limit carbon dioxide emissions only showed a case for very modest cuts, about 11 per cent.[40] Although a Framework Convention on Climate Change was adopted at Rio, it was modified under American pressure to omit any quantitative targets for the reduction of greenhouse gas emissions. The search for a more substantial treaty continues.

In a world of states, responsible for discrete territories, governments distinguish between territories that they are, or are not responsible for. 'Here' is anywhere within the jurisdiction of the state. 'There' is anywhere outside the jurisdiction of this state (including the global commons, beyond *any* one state's jurisdiction). It is also a near-universal observation that we value the present over the future; that we 'discount' the future. Why else do we expect a rate of interest for money lent? £100 now is preferable to £100 in a year's time. Either we or the lender may be dead in twelve months. But £110 in a year's time may tempt us to postpone pleasure now, for more later. (This used to be called a sin, i.e. usury. Islamic Sha'ria still forbids the practice and in Sau'di Arabia, Iran and Pakistan commercial banks go to elaborate lengths to disguise giving or receiving interest.)

Table 1.4  Time/distance and priority given to environmental problems

|          | *Here*   | *There*  |
|----------|----------|----------|
| *Now*    | 1st      | 3rd/2nd? |
| *Future* | 2nd/3rd? | 4th      |

Threats to the here and now have the greatest salience and priority. As citizens, we may debate whether our second priority concerns threats to others now or threats to ourselves in the future. Almost certainly, threats to other people in the future rank fourth. Although it is fashionable to proclaim the global nature of threats to environmental quality (climate change, ozone depletion, marine pollution, the trade in endangered species, etc.), many environmental threats to human welfare *are* territorially discrete. The fetid slums of the Third World *are* in the Third World. They co-exist with great riches in the same cities. Empathy may be transnational, but sewage is stubbornly local. (Unless, like the UK, you dump tonnes of it in the North Sea, or what Fifers still call the German Ocean, and trust to the helpfully counter-clockwise current of that small sea to transport it to Holland's beaches.) Deforestation and desertification are also national, and sometimes regional but not global processes. Even the refugee problems that they precipitate are overwhelmingly regional in their impact. Most of the refugees of Sudan, Somalia and Ethiopia are contained *within* the Horn of Africa.

*Concern* for others is free, but given the distinction between 'here and now', and 'there and then', it is harder to accept that redistributive taxation, the financial options forgone, and restraints on consumption are not without sacrifice. Action involving costs to the taxpayers of the developed countries is therefore likely to fall well short of those peoples' *un*costed sentiments. At election-time voters weigh their conscience against their wallet, which may partly explain the failure of opinion polls to correctly forecast the UK general election result in April 1992. Labour voting intentions were over-reported beyond the usual and admitted limits of sampling error. Perhaps the Galbraithian 'politics of contentment' was at work. Similar observations on the decline of the public-sector services were made by Albert Hirschman in 1970s.[41] General affluence does not assist the public sector. Rather, rising disposable income allows an increasing number of citizens to purchase superior services from the private sector. For those with the money to exercise choice, a second car, a private pension plan and a quick private operation are more desirable than the bus-queue, the eroded state pension and NHS waiting lists. Of course, the working poor and the underclass of three million unemployed do not have the choice. The welfare system can thereafter be eroded. It becomes politically expendable.

It is no coincidence that the two countries that have done most to advance the neo-conservative agenda of privatisation and 'rolling back the frontiers of the state' have also been the most critical of the United Nations' pretensions to global regulatory competence on questions of environment and development. The Reagan–Thatcher treatment of the UN during the 1980s has been discussed above. The US–UK two-step; withdrawal from UNESCO; the refusal to sign the Montego Bay Convention; and the pursuit of 'zero-budgetary growth for UN agencies throughout a decade of devastation for the Third World poor; these are more revealing of Anglo-Saxon attitudes to sustainable development than any number of 'Darwin Initiatives' funded at £6 million, as announced by Prime Minister Major at Rio.

Yet, markets *can* be harnessed to the cause of sustainable development. Environmental preferences can be expressed through the price mechanism. David Pearce, Michael Grubb and others have shown the way forward. Asked to *pay* for the protection of whales, elephants or trees, conservationists may object that these are priceless, reflecting an absolute moral commitment to defending the defenceless. Unfortunately, as Pearce has shown, 'priceless' means exactly that, if they have *no* price and are therefore 'free' for exploitation, those people who *are* prepared to put a cash-value on them will do so. The market can help in those circumstances when regulation is not feasible or verifiable. The price mechanism can be harnessed to encourage responsible behaviour. Whalers, poachers and loggers can be *bought out*. Anti-whaling and anti-logging NGOs and governments might bid for permits that they would *not* use, against whalers and loggers who *would* use them. Thus they might buy-off those who *have* put a price on the priceless.

In the UK, landowners are paid *not* to plough or to drain sites of special scientific interest (SSSI's) designated on their land. The EC and CAP reforms using 'set-aside' could be adapted for creative environmental-conservation purposes. Asking conservationists to outbid those who want to eat what conservationists wish to save, may sound ruthless, but domestic and EC precedents exist. It may be the moral high-ground to claim that some things are priceless, but the real-world question asks, 'What is the price that populations are prepared to pay to maintain the environmental values they wish to preserve?' If electorates wish to include environmental security within their wider understanding of security, then, as with different defence policies, different political parties must offer different, fully-costed programmes and policies. *Agenda 21*, a remarkable document in many ways, is especially remarkable for costing the forty substantive issues it addresses, at the staggering total of $600 billion *per annum* between 1993 and 2000.[42]

This study is focused upon the role and limitations of the United Nations as a forum for environmental diplomacy, with particular attention to UNEP and UNCED. It will therefore make many assumptions concerning the utility of international organisations which reflect the pluralist scheme. In so doing it will also seek to honestly explore the limitations associated with that school of thought. Different approaches to international relations theory reflect different political and cultural attitudes to the natural environment. Young has demonstrated the Judeo-Christian basis to 'man's dominion over nature' which, it can be argued, is the common fault of all mainstream theorising from the Realist to Marxist.[43] In one crucial respect the Biblical view is shared by all mainstream approaches. The mainstream endorses an anthropocentric and ultimately materialist attitude to nature, and thus seeks to fashion nature to *human* purposes. Threats to the environment are primarily recognised in terms of their threat to *human* life-support systems. The environment threatens us and so must be tamed, whether in the defensive interests of the *status quo*, the managerial schemes of the liberal-pluralist or to realise the aspirations of the global poor. None within the mainstream operate on the ecological principle that humankind must both find and keep its place *vis-à-vis* the rights of other species, recognising *their* intrinsic ecological value. Dobson makes the distinction between environmentalism and ecologism, that

> ecologism argues that care for the environment (a fundamental characteristic of the ideology in its own right, of course,) presupposes radical changes in our relationship with it, and thus in our mode of social and political life.[44]

While firmly wedded to the notion of enlightened self-interest which pluralism advocates as the only basis for cooperation between sovereign actors such as states, this study addresses the plight of the global commons and seeks to revive the centrality of the common heritage of mankind concept.[45]

Negotiating binding agreements on the use of the natural environment in general and the global commons in particular, recognises limits to human behaviour on managerial, conservationist and stewardship principles. This scheme therefore recognises the limits and failures of markets. The decade of the Reagan–Thatcher consensus in the years 1979–89, did much to expose the limitations of *dirigiste* economic thinking. The final collapse of state socialism also exposed the appalling state of environmental damage in the former Soviet bloc and the failure of central planning which took no account of environmental values. The market also fails. By definition, the price mechanism cannot cope with environmental degradation created by external costs, and by the free-rider problem and transboundary pollution. Public-goods theories are central to environmental management. This has precipitated the search for economic techniques that identify and rectify these market-failings without reviving state-socialism and regulatory intervention. At best, the price-system may even be capable of generating powerful incentives for rational management of natural resources through the internalisation of costs, 'green-taxation' and other novel instruments such as tradeable permits for levels of permitted pollution.[46] These will be explored extensively in later chapters.

While advocating stewardship of the global commons, this study does not constitute an exploration of ecologism, a task performed well elsewhere.[47] Rather, this study accepts the real-world limitations of promoting environmental diplomacy in a post-Cold War world of 189 states. The recession in the OECD region since 1990 and the oil-glut since 1986 have already combined to challenge the brief prominence of the environmental agenda in Europe and North America that was evident at the turn of the decade. This study is primarily concerned to explore the agenda of reform in the UN system so that the world's states can use that system better to address the triple challenges of environmental degradation, debt and instability which this introduction has demonstrated are intimately connected.

# 2 Debt, Poverty and Environment

At the end of every seven years you shall grant a release. And this is the manner of the release: every creditor shall release what he has lent to his neighbour; he shall not exact it of his neighbour, his brother, because the Lord's release has been proclaimed. Of a foreigner you may exact it; but whatever of yours is with your brother your hand shall release.

Deuteronomy 15: 1–3 *Revised Standard Version*

It is the children who are bearing the heaviest burden of debt. And in tragic summary it can be estimated that at least half a million young children have died in the last twelve months as a result of the economic slowing down or reversal of progress in the developing world.

UNICEF, *State of the World's Children Report, 1989*

The burden of external debt carried by the Third World stood, in 1990, at an aggregate figure of $1319 billion.[1] The flow of repayments, of both interest and capital on this burden, has created, since 1982, a net transfer of resources from the South to the North, variously estimated at between $228 billion, on Miller's figures for net transfers 1982–90, or the $418 billion estimated by Susan George for the same period. The latter estimate is the difference between the OECD figure of $927 billion flowing to the developing countries in all forms, *including* aid, that is, Official Development Assistance (ODA), as well as public and private sector loans, and debt servicing of $1345 billion, paid by developing countries in that period.[2] Both Miller and George exclude from their calculations a number of probably unknowable factors such as capital flight and transfer pricing on the part of multinational corporations (that is, overcharging for *internal* transactions to achieve lower tax liabilities in higher-tax jurisdictions). They also exclude an illegal flow of capital which benefits the developing countries, namely dollar flows arising from trade in narcotic drugs. George highlights the magnitude of the net flow from the poor to the rich, in a striking parallel with the much-admired Marshall Plan. As part of its commitment to the postwar reconstruction of Western Europe, the United States government made available $14 billion at 1948 prices and exchange rates. When adjusted for inflation, George estimates that the Marshall Plan represented a transfer of $70 billion in 1991 prices. Third World debt-servicing has therefore transferred six Marshall Plans from the poor debtors to the rich creditors.[3]

This burden of debt is very unevenly distributed, with a particularly heavy part of it falling upon sub-Saharan Africa, at $160 billion, and Latin America at approximately $480 billion. Miller further divides the nature of long-term debt roughly half-and-half between debts owed to official sources, $517 billion and private banks, $522 billion.[4] Northern attention focused upon the issue after the threatened default of the Mexican government in August 1982, closely followed by that of the Brazilians. Defining the debt crisis in terms of the risks to northern private banks, and claiming the end of the crisis, when in Autumn 1992, that particular anxiety began to recede, are misleading, and serve to mask a vital linkage between the environmental and developmental agendas.

Apart from the *impropriety* of default, interest in the debt issue focused on the instability of the banking system in the developed countries. These so-called sovereign debts (to distinguish loans to governments from loans to private institutions), are carried as an asset on the bank's balance sheet. In the event of default a proportion of this risk would have to be written off as bad debt, and provisions would have to be made to cover that eventuality from the pre-tax profits of the bank. If the bank's creditors feared that their deposits were in practice unrecoverable, this might precipitate a banking crisis as depositors sought to recover their deposits, before the bank was forced to declare bankruptcy. As in the wave of bank failures in the American mid-west during the 1930s, such a 'run on the bank' can become a self-fulfilling prophecy, as creditors scramble to liquidate their holdings, forcing the banks to call in 'good' as well as 'bad' debts. British commercial banks certainly exploited their depositors' fears, to justify higher interest rates, and provisions against bad debts, all in the name of protecting their longer-term viability.

## THE ENVIRONMENTAL IMPACT

Public knowledge that a net transfer of resources from developing to developed countries actually reverses the aid-flow, and the impact this has on the quality of the Third World environment has been relatively weak. The human impact of the debt-repayment burden is catastrophic. In brief, the economic gains built up by the developing countries during twenty years after 1960 began to go into reverse during the 1980s. Per-capita income in sub-Saharan Africa actually went into decline throughout the 1980s. Since so many Africans live so much of their lives outside the money economy, the truer measure of the impact of this deterioration is recorded in indicators such as infant mortality, weight–height tables, life expectancy, etc.

In the developing world, 12.9 million children under age 5, more than 35,000 a day died in 1990 of diseases, most of which were once as common in developed countries. In other words, these children are dying of diseases for which effective prevention, as well as effective treatments, are available.[5]

Of course, not all of these medically preventable deaths can be attributed to the debt burden. However, Richard Jolly of UNICEF traces the deteriorating condition of Third World health expenditures during the 1980s. Thirty-seven of the world's poorest countries halved their expenditure on public health and reduced educational expenditure by up to 25 per cent in this time.

For almost nine hundred million people, approximately one sixth of mankind, the march of human progress has now become a retreat. In many nations development is being thrown into reverse. After many decades of steady economic advance, large areas of the world are sliding backwards into poverty... . It is children who are bearing the heaviest burden of debt. And, in tragic summary it can be estimated that half a million young children have died in the last twelve months as a result of the slowing down or reversal of progress in the developing world.[6]

The most common causes of child-death in developing countries are diarrhoea (3.2 million) and respiratory infections (4.3 million), predominantly pneumonia, but also whooping cough. Both are lifestyle diseases, typical of poor water quality, sanitation and undernourishment, in the way that certain First World diseases, such as smoking-related cancers and cardio-vascular diseases, are also associated with patterns of consumption. The difference lies in the lack of choice that confronts people afflicted with the diseases of poverty and environmental pollution rather than those of the good life. Infant health is inextricably linked to maternal health:

Exclusive breast-feeding in the first six months of a child's life, for example, can dramatically reduce the incidence of diarrhoea; the addition of even water or tea to the infant's diet has been found to double or sometimes treble the likelihood of diarrhoea.[7]

Vaccine-preventable diseases accounted for 2.1 million child deaths in 1990, despite the efforts of immunisation programmes promoted by WHO and UNICEF during the last 15 years. In addition to the well-publicised programme for the elimination of smallpox, parallel programmes have targeted measles, pertussis (whooping cough), tetanus, polio and tuberculosis. WRI estimates an 80 per cent average take-up of these programmes. Malaria cannot be combated by vaccine. Its containment depends upon overall strategies of public health care, and is in fact exacerbated by several development strategies such as irrigation, hydro-electric schemes and other sources of impounded stagnant water. Malnutrition affects one-third of children in LDCs, or approximately 150 million children. UNICEF's category of severe malnourishment is applied to 25 million of them. UNICEF attributes malnutrition as a factor in one-third of all child deaths. Maternal anemia is a critical factor in low-birthweight babies, who are more vulnerable to infection. Among survivors, vitamin A deficiency alone is responsible for 250 000 permanent cases of blindness in children per annum.[8]

It is not only the children of the poor in the Third World who pay the ultimate price for their parents' government's indebtedness. Poverty, compounded by the burden of debt repayment, directly impacts upon numerous aspects of environmental quality. The expression, 'quality of life' does, after all, presume that citizens survive childhood. Adult health quality suffers in levels of air and water pollution, which reflect the rush for a variety of export-led growth which pays the least attention to environmental standards. Spectacular cases of pollution accidents such as Bhopal are exceptional.[9] The *routine* operations of many industrial processes directly affect the quality of life and life expectancy of workers and residents alike. Weir cites particular instances of Third World pesticide producers in Mexico, Taiwan, Tanzania and Indonesia using outdated practices and lacking an 'industrial culture' oriented towards safety controls and management practices.[10]

The need for many indebted countries to maximise export earnings, as part of their IMF-led programme of structural adjustment, in some cases directly conflicts with the lifestyles and survival of traditional peoples within those territories. The impact of deforestation and gold-mining on the Brazilian Indian populations, of commercial logging on the Arawak peoples of both Malaysian and Indonesian parts of Borneo, and the fate of the pygmy people in the stands of tropical rainforest remaining in Zaire and Cameroon are, in essence, questions of human rights. Traditional peoples often with no concept of private land titles, may find themselves subject to a process of internal colonisation as ruthless as that perpetrated by the original European colonial power in the last century. Understandably the governments concerned prefer to view the issue as one of sovereignty. The language associated with the so-called new international economic order (NIEO), dating from the 1970s, encourages this attitude. As will be shown, both the Stockholm and Rio conferences adopted declarations which, although compelling states to recognise their obligations to control transboundary pollution, also enshrined sovereign control over natural resources.[11] Some very bitter exchanges took place on indigenous peoples' rights during the negotiations on forestry questions during the 1990–92 preparations for the Rio UNCED. Discussion of the 'Guiding Principles for a Consensus on Forests' triggered exchange on item 13 , which advised the parties to 'respect the rights and interests of forest dwellers, local peoples and indigenous communities who rely on forests to maintain their livelihood, social organization and/or cultural identity'.[12] The *Agenda 21* document arising from Rio did more to specifically mobilise indigenous peoples in adopting national plans for sustainable development.[13]

Susan George makes three very specific charges, concerning the impact of debt burdens on rates of deforestation, the closely-related assault upon reserves of bio-diversity and, thirdly, the transboundary effects of localised industrial pollution resulting from attempts to locate dirty industries in the Third World but immediately adjacent to First World markets. George describes the plight of the *maquilladora* on the US–Mexico border, rare in being a direct land frontier between the developed and developing world. (Europe's Third World frontier is the Mediterranean, and Turkey its buffer, which shades imperceptibly from the near-European standards of living within sight of its Aegean coast to the poverty of its Kurdish mountains 1000 miles to the east. Japan and Australasia are in turn separated by the China Sea and the Straits of Torres from direct contact.) One official US government estimate has suggested a figure of $9 billion to clean up the environmental damage sustained in terms of heavy metals and untreated sewage pollution in the *maquilladora* zone.[14]

Concerning deforestation, George makes extensive use of data from Norman Myers' numerous publications on the rates of tropical deforestation, and seeks to correlate Myers' data with other sources on indebtedness.[15] The correlation is impressive in establishing that those countries practising high rates of deforestation are also among the most indebted. Seven of the top ten debtors are countries which support tropical forestry: Brazil, Mexico, India, Indonesia, Nigeria, Venezuela and Philippines. Whereas Argentina, China and South Korea are classified as non-tropical timber countries. George chooses to exclude Israel, Egypt, Turkey and Eastern Europe from her table of the most indebted. The former three are excluded on the arbitrary grounds that their debts are primarily derived from military-equipment purchases. Anyway, they are not rainforest territories. Using data from both Myers and from the World Resources Institute, George then ranks the largest deforesters of the 1980s, and summarises her findings thus:

> Of the 24 largest debtors, 8 never had, or no longer have forest reserves significant on a world scale. Of the 16 remaining major debtors ($10 b. or more), *all are found to be on the list of major deforesters.* The correlation is particularly strong for mega-debtors such as Brazil, India, Indonesia, Mexico, Nigeria (among the top 8 debtors) all of which rank among the top 10 deforesters according to both the World Resources Institute and Myers.[16]

Myers acknowledges the role of indebtedness in encouraging deforestation, but is more cautious in his attribution. He cites population growth, unequal distribution of land tenure, and the general neglect of subsistence agriculture on development planning. He also cites the growth of urban unemployment as a factor in closing alternatives for the 'shifted cultivator'.

> Throughout the 1980s the deforestation focus has been almost entirely confined to the logger and commercial rancher. We do not even have a basic idea of how numerous the shifted cultivators have become; we have only broad estimates that range from 200 to 500 million. If the latter figure is correct, they account for almost one in ten of humankind. Yet as margin dwellers they remain a largely forgotten phenomenon.[17]

Both Myers and George agree that rates of population growth, although high, have been massively exceeded by the rates of deforestation over the decade of the 1980s. The rate of deforestation is itself an ambiguous indicator. Measured as a percentage of remaining forest-cover, Brazil's modest 2.3 per cent annual deforestation masks the fact that the area concerned, approximately 50 000 sq. km, is an area equal to the combined annual destruction of the next 10 countries combined. Whereas the apparently declining *rate* of deforestation in, say Côte d'Ivoire, disguises the fact that with 90 per cent of the original forest cover already destroyed by 1989, there is little left to remove.[18]

Table 2.1   Deforestation and population comparisons, 1989

| Country | Rate of deforestation | Population growth |
|---------|----------------------|-------------------|
| Cote d'Ivoire | 15.6 | 3.7 |
| Nigeria | 14.3 | 2.9 |
| Thailand | 8.4 | 1.5 |
| Madagascar | 8.3 | 3.2 |
| Vietnam | 5.8 | 2.5 |
| Philippines | 5.4 | 2.6 |
| Mexico | 4.2 | 2.4 |
| Central America | 3.7 | 2.8 |
| India | 2.4 | 1.8 |

*Sources:* first column, George, op. cit. , 1992, p. 12; second column, Myers, op. cit., 1992, p. 446.

These figures highlight the predominance of African countries among the highest rates of both annual deforestation and population growth rates. Côte d'Ivoire, Nigeria and Madagascar are in the top four places in both lists respectively. Cote d'Ivoire's population growth rate of 3.7 per cent per annum will more than double the population of the country in 21 years. A final year postgraduate student, aged 24, would graduate, and join a population that had grown on an index from 100 to 230 in the time since his or her birth.

To a large extent the threat to biological diversity is synonymous with the threat to tropical rainforests, as the most publicised threats to exotic species and possible destruction of so-called 'Vavilov centres', that is pools of great genetic variety, centre on the forests. There are, however, other sensitive eco-systems at risk from rates of fast and dirty industrialisation, driven by indebtedness to maximise returns and minimise environmental controls. Mangrove swamps, other wetland, inshore fisheries, and pollution of closed seas are all vulnerable.

The Central American Vavilov centre has given the world the original germ-plasm for corn, beans, tomatoes, cassava, squash, gourds, sweet pepper, upland cotton, sisal and cacao. As all plant breeders know, the seed varieties farmers buy are only a step or two ahead of their predators. To keep that lead, ensure resistance to disease and pests, and continue to increase yields a constant supply of fresh genetic material is essential.[19]

The preceding discussion has shown how poverty compounded by indebtedness directly impacts upon the environmental quality of the Third World. This causes personal tragedy and enormous waste of human potential at the most intimate and local levels. It also creates effects at the levels which more usually engage western interests and concern. As Morris and George and Myers show, issues of acute *global* concern, such as deforestation, greenhouse-gas emissions and loss of biological diversity are in part attributable to, and always exacerbated by the complications of indebtedness.

### THE ORIGINS OF THE DEBT BURDEN

How did the situation arise in which the developed countries became the beneficiaries of aid in reverse? The flow of resources not only inverts the logic of ODA, but also serves to stymie popular prejudices concerning the alleged burden of aid. The flow also totally dwarfs the United Nation's two-year budget, currently a little over $2 billion, which has itself been the subject of so much negative publicity in the name of zero-budgetary-growth since 1985.[20] The rise of the debt problem since 1979 is particularly ironic in view of the claims of the so-called New International Economic Order (NIEO). What was anticipated might be a decade of unprecedented attention to the development needs of the world's poorest people turned instead into a decade of outright reversal of growth for many, and the dissipation of the tenuous development gains of the preceding two decades.

The NIEO set out in a series of UN resolutions, a programme of action for the 1970s and 1980s. A series of resolutions adopted in the General Assembly on 1 May 1974 addressed the reform of the international economic system, 'based on equity, sovereignty, interdependence, common-interest and cooperation among states...' and sought to 'correct inequalities and redress existing injustices, [and] make it possible to eliminate the widening gap between developed and developing countries'. The NIEO addressed five basic issues of international political economy.

- Trade reform, especially the stabilisation of export earnings, in a volatile, declining market for commodities.
- Monetary reforms, centrally concerning the Bretton Woods institutions IMF/IBRD (World Bank), their loans policy, and conditions imposed on borrowers.

- Industrial development, with specific reference to technology transfer, and assistance with best available technology, therefore allowing for higher value-added which normally accrued to OECD countries.
- Sovereignty over resources and regulatory powers over transnational corporations. The NIEO resolutions were backed by a 'Charter on the Economic Rights and Duties of States', which, as in the Stockholm Declaration, expressly affirmed sovereignty over natural resources, and added the state's right to regulate foreign investment, and right to nationalisation of assets with compensation.
- Aid: the NIEO called for the OECD to honour the 1970 United Nations pledge of 0.7 per cent of GDP for ODA. This pledge alone, if implemented in 1991 would have *doubled* official development assistance from $55.7 billion total to a figure of over $110 billion.

The expectations of the developing countries were, to say the least, optimistic. However, the assumptions in the developed countries were also profoundly optimistic at this time, and contributed to the conditions which created the debt-burden some ten years later. The use of the OPEC surplus monies derived from the first oil-shock of October 1974, for so-called recycling to the non-oil exporting developing countries, did not so much solve the problem but rather created it. Lenders wanted to lend, and borrowers wanted to borrow. The optimism of both sides was based on a number of assumptions, that collectively presumed the continuation, as normal, of several, actually *exceptional* features of the two decades of postwar growth in the world economy from 1950 to 1970. These assumptions may be summarised thus:

- The continuing fast growth of the OECD region economies.
- Firm, rising commodity prices for Third World raw materials.
- The continued reduction in tariff barriers.
- The maintenance of low, real interest rates.
- A transfer of capital, both investment and official aid, from North to South.

All five assumptions concerning the course of the world economy were to be falsified during the next twenty years, 1973–92, creating the conditions for the debt burden with its human and environmental consequences. In brief summary, it should be noted that during the period of slower growth after 1973, and the actual recessions of 1975–78, of 1981–84, and that since 1990, the growth in the developed countries' imports slowed, and the prices paid for raw materials slumped. This was compounded by the extensive reimposition of protectionist barriers, in spite of the rhetoric of successive GATT rounds. Perversely, slower growth and periods of recession were accompanied by high rates of inflation, creating so called stagflation, a combination of recession and inflation thought impossible by both neoclassical and Keynesian economists. When brought under control, in the Reagan–Thatcher style of the middle 1980s, by tight monetary policy, this created historically high nominal *and* real interest rates. (During the early phases of the oil-shock, despite high *nominal* interest rates, *real* rates were in fact very low, or actually negative, as retail-price inflation in the UK and US for example, went over 20 per cent in the 1975–76 period.) Finally, the commitment on the part of the OECD countries to raise the proportion of GDP paid as ODA to the Third World, the so-called 0.7 per cent target, was not implemented except by a handful of countries. In fact the percentage of GDP offered as aid fell consistently throughout the 1980s, to new historic lows on the part of the UK, USA and Italy in particular. The exceptions to this trend were the Scandinavian countries, the Netherlands and certain OPEC countries. Thus a decade of 3 Rs (recession, rescheduling and Ronald Reagan), was compounded by the 3 Ds (declining commodity prices, demographics and debt).

## (i) The Effects of Slower Growth

The slow-down in the growth of the world economy is well demonstrated in broad terms by comparing the averaged, annual growth rates for each decade from 1960. Such large aggregates combine local differences and arguments between social democrats, liberals and conservatives, and demonstrate the across-the-board nature of the economic slowdown of the last 20 years.

Table 2.2    Average annual rates of GDP growth (percentages)

| Region | 1961–73 | 1974–82 | 1983–90 |
| --- | --- | --- | --- |
| DMEs | 5.0 | 2.1 | 3.2 |
| Africa | 4.9 | 1.2 | 0.4 |
| L. America | 5.6 | 3.2 | 1.8 |
| S. and E. Asia | 5.6 | 5.9 | 6.0 |
| (at constant 1980s prices and exchange rates) | | | |

DME = developed market economies
*Source: Global Outlook 2000*, United Nations, 1990, p. 32.

When these figures are deflated by the effects of population growth, so as to show growth of GNP per capita, the differential impact of this slowdown is even more apparent.

Table 2.3    Average annual growth of GDP per capita (percentages)

| Region | 1960s | 1970s | 1980s |
|---|---|---|---|
| DMEs | 3.9 | 2.4 | 2.1 |
| China | 2.0 | 4.1 | 7.5 |
| N. Africa | 8.2 | 1.3 | –0.3 |
| Africa | 1.8 | –0.4 | –2.6 |
| W. Asia | 4.1 | 1.0 | –4.3 |
| S. and E. Asia | 2.6 | 4.1 | 3.7 |
| L. America | 2.7 | 2.4 | –1.1 |

(DME = developed market economies)
*Source: Global Outlook 2000*, p. 10.

The slower rates of growth in the developed countries were in themselves bad news for the export prospects of the Third World economies. However, it should be noted that Third World exports only account for 18 per cent of world exports in the first place. But when the impact of continued population growth rates of 2–3 per cent are added to the equation, and are in fact higher than GDP growth rates, by definition, negative growth, i.e. declining real income, is the result. This has been the lot of the African and Latin American regions for most of the last decade.

**(ii) Commodity Trade**

The twenty-year slowdown in economic growth reduced demand for primary materials and so reduced their prices. This situation was compounded by an unprecedented wave of substitutions and a secular shift away from heavy manufacturing on the part of those sectors of the developed countries' economies that did enjoy faster growth. The declining significance of jute and textiles, steelmaking, shipbuilding, and auto-assembly, served to reduce demand for many Third World staples. Very high levels of agricultural protection exacerbated the problem, as the EC, in particular, created surpluses of sugar and vegetable oil, which not only reduced demand from traditional tropical *exporters*, but EC surpluses were also dumped on the domestic market within the developing countries, distorting their local markets.

At the root of the commodity price problem is the inelasticity of both the supply of and demand for primary products. Small movements in either supply or demand produce disproportionately large movements in prices.

A cyclical boom in the developed countries strengthens demand for basic materials, especially metals, and thus leads to big increases in the value of primary product exports. Conversely, world recession leads to a sharp fall in the value of exports of primary products, and therefore in the export receipts of the typical developing country.[21]

Miller plots the 'long downward march' of real commodity prices, deflated by the price of manufactures, over the period 1870–1986. Using the 1980s as an index of 100, he calculates that real commodity prices fell from a 1950 high of 131 to a 1986 low of 69.[22] Even the infamous exception of oil prices has been in reverse since 1986, with the posted price falling from a 1979 high of $32 per barrel at the time of the Iranian revolution, to a 1986 low of $11. The price stabilised in the early 1990s at about $17, creating the cheap-oil glut that has done so much to undermine pressure on all oil-consumers, developed and developing alike, to address the energy-efficiency implications of the climate-change question.

The rise of new materials (polymers, carbon-fibre and composites), new technologies (micro-chips, lasers and cell-culture), and whole new sectors of economic activity (information technology, bio-technology, leisure and financial services), has not been based upon the old notion of vertical integration between Northern manufacturer and Southern raw materials. It has created an economic system simultaneously, and ironically, more internationally interdependent than at any time in history, but predominantly reflecting an interdependence of the rich countries upon trade with each other. EC–Japan–US trade is over 80 per cent of the world trade total, whereas fifteen years ago a British university campus would have been home to young men sporting clothes and cultural artifacts from India, Africa and beyond. The change in popular culture is so profoundly advanced that most young men are happy to dress like 1950s Americans for leisure (501s, white T-shirt and baseball hat), and to then graduate dressed as 1960s Americans (button-down shirt, narrow lapel suit and hair-gel). Women are less frequently fashion victims, or perhaps gingham frocks just look stupid.

## (iii) Protectionism

Compounding the effects of slower growth and reduced demand for Third World primary products, the surreptitious rise of protectionism through these two decades also damaged the prospects for Third World economic growth. Although the developed countries were nominally committed to trade liberalisation through GATT, Williamson observes five basic shortcomings in that system. A number of exporters are subjected to so-called voluntary export restraint. Although most well-known in the case of Japanese automobile exports, numerous Asian newly-industrialised countries (NIC) are subject to the same veiled threats, especially in textiles. The multi-fibre agreement is 'a multi-lateral agreement which allows unilateral action to limit quantitatively all forms of textile imports. It is used to limit exports from the developing countries to the developed countries.'[23] Second, most trade in agricultural products falls outside GATT regulations. Thirdly, GATT has allowed blanket application of the so-called infant-industry protection within Third World members, discouraging modernisation. Fourthly, the GATT rules permit countries to apply quantitative restrictions on imports with IMF permission. Finally, GATT permits the formation of regional trading blocs, such as the EC, offering discriminatory preferences between members.[24] The proportion of world trade subject to so-called non-tariff barriers has greatly expanded during the last twenty years.

Table 2.4   Percentage of major imports facing non-tariff barriers, 1965–86

| Importer | Food 1965 | 1986 | Manufactures 1965 | 1986 | Agricultural (raw) 1965 | 1986 |
|----------|-----------|------|-------------------|------|-------------------------|------|
| EC | 38 | 96 | 2 | 56 | 15 | 58 |
| USA | 17 | 57 | 27 | 59 | 5 | 46 |
| Japan | 51 | 99 | 11 | 41 | 2 | 69 |

*Source:* Miller, op. cit. , p. 49.

## (iv) Real Interest Rates

Using the statistical tables of the Bank for International Settlements, Miller argues that 'the historical average of real interest rates' has been in the range 1 to 2 per cent. During the period 1986–91, the level has averaged at

6.8 per cent.[25] The source of these historically anomalous rates was the so-called Volcker shock, named after the Chairman of the Federal Reserve Bank, who, in 1979, initiated the rise in American interest rates. The need for this policy arose in part from the *other* debt crisis: namely, the expansion of US government public debt during the same period. When the aggregate of Third World public debt stood at approximately $1300 billion in 1990, the US federal government debt stood at $675 billion, a figure five times larger than Brazil's foreign debt.[26] Despite nearly a decade of balanced-budget rhetoric and the disciplines of the Gramm–Rudman–Hollings Act, it was estimated that, by 1992, approximately 15 cents of each federal tax dollar raised was paying interest on the deficit. By the close of the Reagan presidency the federal deficit had in fact been doubled over the eight years. At current prices Reagan borrowed more than all of his predecessors combined. In order to finance the double-deficit, namely the federal deficit *and* the balance of payment deficit, the US had to set internationally competitive interest rates to attract inward investment and short-term financing. The US can attract foreigners to finance its deficit in its own currency, which gives the US less incentive than many countries to worry about currency depreciation. Other deficit countries, such as the UK, have been obliged to match, or better US rates, and in the years 1989–92 to shadow, or offer more than German rates as well. As Germany sought to finance a burgeoning deficit brought about by the costs of unification, Kohl's preference was to raise interest rates rather than to raise taxes. These stresses were to lead to Italy and the UK leaving the European Exchange Rate Mechanism (ERM), in the spectacular collapse of sterling in September 1992. The whole ERM effectively collapsed in the summer of 1993, reverting to a wide variation in permitted exchange rates. In the supposedly balanced-budget Reagan era, nominal rates climbed to over 20 per cent. These rates endured through the early and mid-1980s, before falling during the early 1990s, but, in that period, the increased repayments required from major debtors triggered the first threatened defaults, by Mexico and Brazil, in 1982. Miller claims that for each increment of 1 per cent in interest rates, the cost to the largest 17 debtor nations was $8 billion per annum.[27]

### (v)  The Myth of Aid

The impact of the four factors cited above might have been mitigated by the flow of aid, or official development assistance (ODA), from North to South. In fact the flow of ODA has fallen far short of the OECD's self-proclaimed targets throughout the two decades under discussion. Although the current dollar value rose, even through the tightest phase of the 1980s, from $36.8 billion in 1981 to $51.7 billion in 1988, the real, inflation-adjusted, amount fell. In percentage terms ODA represented 0.34 per cent of the OECD countries' GDP in 1991. Contrary to US domestic apprehensions, Japan is the largest aggregate donor at $11 billion. (US figures include arms purchase credits in the overall 'foreign assistance' figures.) In per capita terms, and as a percentage of GDP, American ODA of $7.7 billion in 1989 was the lowest of the 21 aid-giving members of the OECD.[28] Multilateral disbursements run at approximately four-fifths of the total of ODA. Within the multilateral amounts are the role of the World Bank and its soft-loan associate, the IDA. In its dealings with the highly indebted countries the World Bank has in its turn become a net *recipient* of Third World repayments, recording a net transfer of $ 1.267 billion in 1988 and $1.926 billion in 1989.[29]

Williamson, with a longer historical perspective, charts the decline of ODA as a percentage of OECD countries' GDP from as early as 1960 (0.51 per cent) to the 1980s (0.37 per cent ). The only countries to be meeting the 0.7 per cent target have been Denmark, Netherlands, Norway and Sweden among the OECD. Frequently overlooked have been the conservative Gulf states, with Kuwait, Qatar, and UAE donating *over* 3 per cent of GNP. Williamson makes the environmentally fascinating observation that for the Gulf States in particular, and the OPEC in general, basing so-called GNP figures upon extracting a non-renewable resource is anyway misleading.

> Conceptually correct (though not currently conventional) accounting would treat the depletion of oil reserves in the ground on a par with the depletion of produced capital goods. This would mean that the overwhelming part of oil revenue would not qualify as part of net domestic product, although it does of course contribute to GDP.[30]

The compounded effects of slower growth in the North, declining commodity prices and incipient protectionism slowed growth rates throughout the poorest countries of the South. Their predicament as debtors was then compounded throughout the 1980s by rising real interest rates and the falling real value of ODA. The countries to benefit from export-led growth, challenging and then defeating the EC, Japan and the US in traditional 'metal-bashing' sectors were the so-called NICs, a small number of mostly South East Asian, and some Latin American countries. However, despite the impressive performance of these countries, measured by conventional indicators, they also paid a high environmental cost for their preferred pattern of export-led growth. In 1986, Taiwan, South Korea, Hong Kong, Singapore and Brazil accounted for an incredible 55 per cent of all Third

World manufacturing, 46 per cent of Third World manufactured exports and 17 per cent of Third World GDP.[31] Thailand, Chile, Indonesia and Mexico formed a second division of NICs. The environmental cost of this pattern of growth has been substantial.

> Less than 1 percent of human waste receives primary sewage treatment. Probably as a result, the island [Taiwan] has the highest incidence of hepatitis B in the world. Taiwan is one of the top users of pesticides and fertilizers per hectare in the world, and this load contributes to the contamination of surface water and groundwater. Emissions of nitrogen oxides from motor vehicles in Taiwan tripled between 1977 and 1985, and may double again by the end of this decade. Pollution is bad enough to make the air hazardous to breathe on 62 days a year. Asthma cases among children have quadrupled in the last decade.[32]

The common characteristics of the most successful East Asian NICs are curious. They are for the most part authoritarian in their politics, yet relatively efficient in public administration. They are all anti-socialist and yet corporatist, enabling intervention by government in industrial policy. Although competitive rather than egalitarian in social attitudes, they are committed to improving social infrastructure with particular attention to high levels of educational attainment. In short, something of the Confucian, rather than Protestant work-ethic abounds in an industrial culture which in many ways matches the confidence in technology and faith in the future that was characteristic of UK and Germany in the 1890s.

## RELIEF, RESCHEDULING AND RETIREMENT

Third World indebtedness worsened during the middle to late 1980s, and had numerous negative impacts besides the most obvious ones discussed so far, namely impoverishment and environmental damage, principally affecting the debtors. George elaborates a number of 'boomerangs' by which negative aspects of the debt burden came back to the developed countries, which were now in receipt of a net transfer of resources. She describes the rise of narcotic drug manufacture and export in countries such as Colombia and Pakistan, supplying the market in the developed countries, especially the US and western Europe. She reveals the costs transferred to taxpayers in the developed countries as banks enjoy tax-write-offs when making provisions for bad debt. George also explains the impact on the export industries of the developed countries that have traditional markets in Third World countries, which reduce their imports due to the commitment of scarce foreign reserves to debt-servicing. George's final 'boomerangs' refer to the impact of illegal immigration on the North and deteriorating military security associated with impoverishment of the debtor countries. These issues have been discussed in Chapter 1.[33]

This analysis suggests a substantial case for debt relief based upon the enlightened self-interest of electors, if not the banking community, in the OECD countries. Many imaginative plans to ease the debt crisis have been evolved during the period since 1985. Miller summarises over 50 of them.[34] Common to nearly all of these numerous plans is the attempt to combine three elements:

- debt relief, i.e. to reduce interest rates and or to reschedule the loan period,
- debt swaps, i.e. to exchange the debt for some other asset,
- debt forgiveness, i.e. cancellation or write-off.

For example, most plans for debt relief include an element of write-off or forgiveness of debt, whereas re-scheduling over a very long period of time may involve converting some part of the debt to a more secure, but lower yield bond. Common to all of these plans has been a marked reluctance on the part of the creditors to accept any *transfer of risk*, such as in government guarantees of loans, and a reluctance to use the vocabulary of forgiveness, even when it has in fact been implemented. It is also significant that throughout the 1980s, hampered by its double deficit on the balance of payments and the federal government, the USA was not in a position to refinance Third World debts with additional liquidity. Only Japan and, until 1990, Germany have been in a position to add additional resources. Within the creditor group of developed countries, there has been a subdued, and sometimes public argument in which Japan and Germany have gritted their teeth in the face of American calls for interest-rate reductions, in a decade in which it has been the collapse of the American domestic savings and political reluctance to raise taxes to tackle the double deficit, that has inflicted the need for such historically high rates of interest.

Strategies to confront the debt crisis therefore range from doing nothing, and so letting the market cope with defaults and other instabilities, to the outright forgiveness of debt. Rescheduling has already been applied on a massive scale, but as lengthening the period of debt will eventually raise rather than lower the *aggregate* payment of interest it essentially shifts the repayment problem on to the next generation. Swaps, as will be discussed shortly, can take the form of debt-for-equity swaps, or the widely canvassed but barely utilised debt-for-nature swap which will be the subject of more detailed analysis in Chapter 6. Miller also discusses 'debt-for-debt' in which debts are converted into long-term bonds, which is in effect, the ultimate rescheduling.[35]

A widely promoted and comprehensive proposal, the Brady Plan, combined varieties of rescheduling, swaps and write-off. It offered a degree of debt write-off, converting the debts to more realistic longer-term loans, with US permission for the IMF to raise its quotas, with additional funding being supplied by Japan. The World Bank and IMF would make more finance available for debt servicing, which would be guaranteed by the US. This last element, in which a public institution assumed some part of the risks historically incurred by private banks, was what made the plan of 9 March 1989 a breakthrough.[36] In practice the Brady Plan proved a disappointment.

Negotiations under the aegis of the Brady Plan reduced Mexico's debt to $92 billion in early 1990, thereby cutting its $12 billion annual debt-servicing obligations to commercial bank creditors by a 'disappointing' $4 billion, *but providing little of the hoped-for new money* that is desperately needed in light of Mexico's dismal trade balance prospects. Offered three options, half of the creditors opted for converting part of their debts into new guaranteed bonds with yields of 6.5%, 40% of them chose conversion to saleable guaranteed bonds worth 65% of their face value, and 10% preferred to provide new money worth 25% of their exposure on condition that they could get guarantees on their old debts.[37]

All this occurred in a period in which real wages in Mexico declined for eight consecutive years, 1982–90, to a real value approximately half that of 1982.

Debt for equity conversions proved briefly attractive during the 1980s as part of the vogue for privatisation. They were used in Turkey, Nigeria, Chile, Mexico and Brazil. Political stability is a prerequisite to attract this sort of conversion. As Miller observes, 'Chile promised stability under dictatorial rule.'[38] Equity swaps were widely blamed, *within* Brazil in particular, for worsening the inflation experienced in that country. Any swap which involves converting a foreign-currency debt into a local-currency instrument will necessarily increase the money-supply within that country and so exacerbate inflation. Among the more exotic varieties of swaps are those arranged *between* indebted countries such as Mexico and Brazil to the value of $100 million and, of direct interest to environmentalists, the 'debt for good-works' swap pioneered by the Midland Bank which donated some probably irrecoverable low-value loans to UNICEF, for conversion to local Sudanese currency, for water-well drilling projects. The total value of all varieties of swaps is estimated at just \$30 billion.[39] It also raises fundamental political questions concerning recolonisation of the debtor countries' assets.

The outright forgiveness of debt has been resisted by the creditors on grounds of precedent, but as will be shown, there are in fact *numerous* precedents for debt-forgiveness. Most obviously previous cases have benefited European countries, as with German reparations write-offs in the Dawes Plan, 1924–29, and again, after the Second World War, when German Marshall Aid debts were largely written-off. More recently, in 1991, Poland and Egypt were the beneficiaries of explicitly politically motivated debt-forgiveness. The first occurred in the context of post-communist democratisation. An American proposal to write-off 70 per cent was limited on Japanese insistence to just 50 per cent. The latter case occurred in the context of post-Gulf War gratitude for Egyptian assistance against Iraq.

For public sector debts, i.e. inter-governmental debt, forgiveness takes the form of retrospectively converting loans to outright grants. For private sector debts, i.e loans to governments by foreign banks, it requires the write-off to be compensated by the bank's home government, above and beyond the public subsidy already implied in tax-liability write-downs. Williamson sounds a cautionary note. He suggests three, closely-related arguments against simple remission of international debts. Firstly, the benefit of cancellation is capricious. It rewards the most indebted, however good or bad the reasons may be for their predicament. Secondly, it creates an incentive for over-borrowing on the part of the debtor; and thirdly, its obvious corollary, a powerful disincentive for creditors to extend further loans.[40] Paul Vallely opens a wholly different ethical perspective, with his emphasis upon both Old and New Testament precedents for the forgiveness and release of debtors. Something similar and secular can be derived from Rawls, or indeed Pareto, in terms of the reasons to effect a bias to the poor in the redistribution of wealth.

Ethics and Biblical precedent aside, the most compelling argument for forgiveness is that the market has *already* given up on a very large part of the debt burden. The very existence of a *secondary market* for discounted debt represents a judgement by the participating banks that they will only ever redeem a very small part of the full face value of the debts they hold. The double standard in this situation is that it is the banks which benefit, through tax-write-downs, from the existence of the discount market, rather than the debtors, whose incapacity to repay is recognised by that market's existence. The extent of the secondary markets's discounting is highly variable.

Table 2.5   Secondary market price of face value of loans (percentages)

| Chile | 89 | Costa Rica | 47 |
|---|---|---|---|
| Colombia | 85 | Nigeria | 42 |
| Jamaica | 72 | Brazil | 34 |
| Uruguay | 62 | Ecuador | 25 |
| Venezuela | 61 | Bolivia | 11 |
| Mexico | 55 | Peru | 7 |
| Philippines | 50 | Côte d'Ivoire | 6 |
| Morocco | 48 | | |

*Source:* for 2nd quarter of 1991 from George, op. cit. , 1992, p. 69 citing data supplied by Saloman Brothers and Merrill Lynch Capital Markets, June 1991.

Miller suggests that recognising the market value of Third World debt would mean, on average, a 60 per cent reduction on face values.[41] Senator Bradley and others have argued that, at least, debt-forgiveness should be extended to the extent that the secondary market already discounts Third World debt. Germany and Sweden both began releasing debtor nations in 1973, followed in 1986 by the Netherlands and Canada in 1987. Within the G7 group, both President Mitterrand and then Foreign Minister John Major proposed a measure of debt-forgiveness in relation to the poorest sub-Saharan African nations in 1987 and 1990. The Major proposal extended to 26 of the poorest countries, at that time owing a total of $116 billion.

Considering that the developed countries are the net beneficiaries of Third World indebtedness, and have themselves benefited from historic acts of debt-forgiveness, the political case for an act of magnanimity towards the Third World debtors would appear to be profound. In view of the environmental damage which is also attendant on perpetuating the present situation, it would appear that debt-forgiveness on the part of the developed countries would also serve the needs of sustainable development. The levels of environmental stress and degradation experienced by many Third World countries beggar comparison with the environmental impacts of affluence associated with the debate in the North. Furthermore, the persistence of the debt crisis and the existence of net transfers of wealth from South to North postpones the ability of the Third World countries to join with the industrialised countries in the joint resolution of genuinely global problems. The flows of money associated with UNEP, UNDP and even the much-vaunted Global Environmental Facility are swamped by the reverse flow of debt-servicing, which as George in particular demonstrates, also create destructive 'boomerangs' for the developed countries themselves. A number of possible ways of increasing the liquidity available to the international system will be examined next, and in Chapter 6. In particular three paths specifically linked to environmental stewardship will be explored, namely:

- revenues, or rents, from the so-called global commons,
- debt-for-nature swaps,
- international carbon-taxes.

None of these is an original suggestion. The use of the common-heritage revenues, in the context of the seabed beyond national jurisdiction was agreed in the United Nations in 1970. Debt-for-nature swaps have been applied on a very limited scale as a palliative for the debt crisis discussed above. Carbon-taxing has been confronted by a number of states, and by the EC Commission, as one tool among many with which to confront rising energy use, and the carbon dioxide emission targets set, or claimed by many OECD countries in their approach to the Framework Convention on Climate Change adopted at Rio in June 1992. However, brought together and synthesising elements of *dirigisme* with a willingness to embrace market forces, they offer a comprehensive way in which to raise liquidity, relieve the burden of debt and fund the expansion of UNEP and UNDP programmes for sustainable development.

# 3 The Global Commons

As was shown in the preceding chapters, the environmental and developmental agendas are linked, not only at the institutional level, as in the UNCED agenda of 1992, or in consequence of the debt burden, but in everyday and obvious ways by the overwhelming impact of mass poverty on the ability of any society to maintain environmental standards that are taken for granted by the industrial powers. The explicit attempt to link the developmental and environmental agenda at UNCED was not new. The linkage reflected an attempt by the Third World countries to revive the debate on development that had failed by the 1990s.

Central to the NIEO was the affirmation of permanent sovereignty over natural resources by countries that had experienced decades, and in some cases centuries of foreign colonial occupation, and the concomitant exploitation of their natural resources for others' gain. However, simultaneously, an apparently contradictory doctrine, namely the common heritage of mankind (CHM), was advanced in parallel with the NIEO debates. Proponents of the common heritage of mankind argued that the wealth of the commons (usually cited as the deep seabed beyond the limits of national jurisdiction, Antarctica and Outer Space), should not be left *res nullius* (Latin: 'an ownerless thing'), for anyone to exploit on the basis of first-come-first-served. Rather, such territories should be taken into common ownership and exploited for the common good, which, in an international system numerically dominated by weak and poor states, meant a commitment to redistribution of wealth, primarily for the benefit of the newly-independent and historically abused developing countries. The approach was well-summarised by US President Johnson, speaking in 1966:

> Under no circumstances, we believe , must we ever allow the prospects of rich harvest of mineral wealth to create a new form of colonial competition among the maritime nations. We must be careful to avoid a race to grab and hold the lands under the high seas. We must ensure that the deep seas and the ocean bottoms are and remain the legacy of all human beings.[1]

The idea has also been soundly denounced, particularly by the Reagan Administration: 'deep sea mining remains a lawful exercise of the freedom of the seas open to all nations'.[2] The tension between sovereignty, *res nullius* and the common heritage of mankind has profound implications for the practice of sustainable development. It is not only possible to make the case for environmental protection by asserting the need to establish a common heritage regime for territories previously left *res nullius*. It is also possible to make the case for the extension of the common heritage concept to territories presently assigned to the exclusively sovereign jurisdiction of states. Thus the common heritage challenges *both* conservative claims to the free use of certain natural resources as well as radical assertions of *state's* rights. This may account for the way in which the common heritage concept was consigned to the realms of outer darkness during the shrillest years of the Reagan–Third-World confrontation.

## ANTECEDENTS OF THE COMMON HERITAGE

The common heritage of mankind is a concept most readily identified with the proto-regime for the seabed beyond the limits of national jurisdiction. In modern times, the idea of the common heritage has postwar precedents in elements of the Antarctic treaty system of 1959. This suspended territorial claims and reserved the continent for peaceful scientific exploration. Also the Outer Space treaty of 1967 similarly prohibited claims to title or territory in outer space and also prohibited the stationing of nuclear weapons in earth orbit or on the moon. The 1967 debates of the UN General Assembly which initiated the United Nations Conference on the Law of the Sea in its IIIrd and IVth rounds of negotiations, 1967–82, were crucial to advancing the idea. The common heritage received near-universal endorsement in General Assembly resolutions adopted in 1970. The concept was also incorporated into the final draft of the Convention on the Law of the Sea, which was concluded in 1982, but which for a complex variety of reasons has not yet entered into force.

More substantial precedents for the concept of the common heritage can be traced to various Roman law traditions of *res communis omnium*, and *res publicae*. Essentially the common heritage idea argues that certain property which belongs to no one, belongs to all, rather than to whoever, first-come-first-served, can establish *use* of that resource. As will be shown, the Roman law arguments show a remarkable resemblance to those elements of public-goods theory discussed in Chapter 1. The legal and philosophical basis of the concept rests upon the need to determine and distinguish between ownership, use and exploitation of goods, services and commodities which by their innate characteristics do not *obviously* fall under rules governing private property. A radical, Thatcherite model of privatisation is possible and will be discussed shortly. In Roman times, rivers, parks and roads were typical of the property which needed to be assigned in this way. Whereas the high seas may be *res nullius*, that is, outside the law, and capable of being taken from, freely by all, the Romans applied different rules to facilities provided for the public benefit. *Res communis omnium*, as public goods, bestowed an indivisible benefit on the citizens. A park or water-fountain, which was used by one person was still available to others. In other words, one person's use did not deprive another of their right to enjoy the same. A problem arises, however, as Pinto explains,

> *Res communis* sufficed so long as there was no exploitative use, so long as no part of the thing was to be taken, transformed or converted by some individual thereby placed permanently outside the reach of the other members of the community who were entitled equally to benefit from it.[3]

Although drinking from a fountain is technically exploitation, like taking one fish from the sea, it may be taken for free, because in practice it does not deprive the next user. More particularly, the distinction between passive enjoyment and commercial use, is typified by public goods such as roads and harbours; that is, *infrastructure*. Such property not only facilitates a wide range of private commercial activities, but it is also subject to depreciation and therefore needs investment and replacement if it is to be of continued use. *Res publicae*, in the Roman view, were 'such things [as] were not, in their original form, capable of individual ownership, but were subject permanently to the collective ownership of the community'.[4] Funding and charging for the provision of such goods has historically rested either upon either general taxation, or, specific charges for use, e.g. tolls. The indivisibility and impracticality of charging for facilities such as streetlighting favours funding them by general taxation, and levying no charge at the point of consumption. Before the invention of the internal combustion engine, roads were sufficiently novel, and the pedestrian or horse-drawn pace sufficiently slow, to allow user-charges to be levied, i.e. tolls, which made such public property self-financing. In time, the provision and use of roads became so widespread that during the early nineteenth century they became 'free'. Thereafter, as motor-vehicle speeds rose from single figures to a legal maximum of 70 m.p.h. in the UK, the practical possibility of collecting tolls receded further. Toll-*bridges* never went out of fashion in Scotland, where Fifers in particular, not only pay to get into the county at one end, but also pay to get out of it at the other end. Tolls levied on both the Tay and Forth bridges were justified on the almost wholly specious grounds that identifiable expenses and benefits arose in the construction of *estuarine* bridges. (If the alternative means of travel to

Dundee or Edinburgh were to swim, the argument for exclusive benefit would hold water. In practice, alternative land-routes exist via the long way round. Any one mile of any motorway is in practice similarly helpful, if not essential. The government's case for a *toll*-bridge to Skye can only be sustained by their closure of the state-owned alternative ferry.) Now, a new political ideology eager to charge for the use of social services, the growth of traffic congestion and the advances of silicon chip technology have all combined to revive the political case for toll roads and sophisticated road-pricing schemes to be applied to city-centres. Whether such charges can also be proposed as environmentally friendly will be discussed later.

The distinction between passive enjoyment and active exploitation was central to the law of the sea debates. For instance, the traditional high-seas right of surface navigation is non-exclusive. However, deep-sea mining of valuable minerals, or oil-exploration on the continental shelf, beyond the limits of national jurisdiction, is, by definition, an attempt to establish exclusive use of that resource. The law of the sea debate sought to preserve the surface of the sea as *res nullius*, and attempted to introduce an element of regulation, on conservation grounds, to fishing rights beyond the limits of national jurisdiction. The debate also sought to establish, as new, the concept of the common heritage of mankind for the mineral wealth of the sea-floor and subsoil, beyond national limits. In this case licensing and tax-ing of mining activities was to be invested in a new UN agency, the still-to-be-created International Sea Bed Authority (ISA), and all maritime law dis-putes were to be resolved through a new legal tribunal to be established at Hamburg.

Berkes demonstrates the existence of a parallel, non-western tradition of communal ownership, which is familiar to anthropologists of traditional societies. Grazing, hunting and water-drawing rights are frequently invested in a tribe, village or commune. No individual within that group has particular rights, neither do outsiders. The property is at one and the same time commonly held by the group and yet exclusive to that group.[5] This incidentally allows for a fundamental challenge to the Hardin thesis. The 'tragedy of the commons', that is, overuse by competing users, is not inevitable, but is based upon western concepts of individualistic economic man. Other cultures do not necessarily rush to exploit common property for private gain.[6]

This suggests that we can identify four possible concepts of property, distinguished not so much by their *intrinsic* nature as by their location, and the user-rights attaching to them. Susan Buck suggests that a salmon may be freely caught *res nullius* at sea but, depending upon whether it then swims up a river in Scotland, or in territory belonging to the US Forestry

Service, or that comprising an American Indian reservation, the unlucky fish would then become, respectively, private property, state property or communal property.[7]

The case for extending the concept of the common heritage of mankind to a number of new territories rests upon several arguments, invoking good management, equity and security. Firstly, there are significant new territories, the seabed, outer space and Antarctica, which are not subject to national territorial claims. To this trio of established commons a case can be made to add the physically more nebulous properties of the atmosphere, including the stratospheric ozone layer, and indeed the world's climate system. These have been treated previously as open-access, free territories, but are now known to be subject to environmental degradation caused by human interference. Regarding them as 'life-support systems' rather than as free dumps means also recognising the need for collective action to maintain their life-supporting qualities. The international negotiations that gave rise to the Vienna Convention and Montreal Protocol on substances harmful to the ozone layer recognise this responsibility. However, obligations and rights cannot exist in the legal vacuum of *res nullius*, and so *some* concept of property, towards which states can *have* obligations or on which they can claim rights becomes necessary.

Secondly, these territories are potentially vulnerable to the so-called 'tragedy of the commons' namely their destruction due to excessive use, so long as each additional individual user has free access to the commons concerned. Emotive arguments can be made that we are 'all in the same boat', often likening the world to a 'lifeboat' supporting the human population on a storm-tossed sea of environmental threats. However, as was shown in the previous chapter, the inequality of international society is so great that some members *can* exhaust non-renewable resources, and *can* over-exploit renewables, *and* can then move on, with sophisticated and substitutable economies, to exploit other resources. This used to be called colonialism. However, the logic of sustainable development is that destruction of the commons violates the obligation of one generation to bequeath a stock of natural capital that does not impair the user-rights of subsequent generations.

A third argument, to advance the application of the common heritage of mankind to new territories, concerns competing claims and military threats. Demilitarisation, which is a characteristic of the Antarctic and outer space regimes, serves to lessen rivalry and tensions over those commons which *are* physically capable of being expropriated, and therefore subjected to disputed claims. For example, one of the factors which encouraged the parties to negotiate on the law of the sea was the tendency towards 'creeping territoriality', whereby many states extended their claims over the high seas and seabed. Some claims extended to 200 miles, others invoked geological evidence to extend claims over the whole adjacent continental-shelf area. The Falklands dispute between Argentina and the UK was complicated by this factor, and Canadian–US relations were strained over the rights of passage by oil-tankers through disputed Arctic waters.[8] It is not technically possible to annex, patrol and police outer space. The common heritage status does not eliminate the question of user-rights, but adopting that approach removes the territories concerned from the arena of military contention, and places them in the *salle diplomatique* of negotiation over quotas, licences and revenues to be paid to the appropriate international authority for their use. The bitterness of the arguments deployed in the UNCLOS sessions were one substitute for war; most wars, however, are much shorter than the twelve years of UNCLOS negotiations of 1970–82.

Here we may re-enter the debate concerning the privatisation of domestic commons, which is analogous to extending state-sovereignty over international commons. The common-land of England was subject to almost total privatisation during the seventeenth and eighteenth centuries. The so-called 'enclosures' simply permitted the local aristocracy and larger farmers to expropriate common-land as their own. More recently, a new vogue for privatisation has arisen, seeking to reverse the consensus on state-ownership established in the 1940s. The boundary between public provision and private ownership was revised very thoroughly in the UK and USA during the 1980s, and a wave of post-communist imitators in Eastern Europe has now joined in that revision. In terms of access to capital there is no reason why the private sector cannot fund massive 'public works' in the same way as the state. Multinational enterprises have gross corporate products that rival smaller European countries' GNP, and dwarf half a hundred members of the UN. In their technical prowess, the number and skills of their staff and probably in their planning time-horizons, these corporations easily outrank the UN system of specialised agencies. The 'can-do' factor is not in doubt. However, private corporations will only enter into the realm of the traditional 'public' sector if assured of a profitable return. This can be achieved in three ways. Each has been used in the UK since 1985.

The private contractor can be paid a subsidy, as recently proposed for loss-making routes of the former British Rail network, now planned for fragmentation into twenty-six franchises. Secondly, the contractor may be granted extensive monopoly powers, as in the privatisation of British energy utilities such as British Gas, the Central Electricity Generating

Board (CEGB) in England, The North of Scotland Hydro Board and South of Scotland Electricity Board, the national electricity grid and the regional electricity distribution companies. Thirdly, the contractor may be allowed to charge user-fees. Fees may be relatively uncontroversial in the case of say, road and bridge tolls. Or they may be bitterly resented if a service which for many years was supplied free at the point of consumption, and financed from general taxation, is suddenly subject to metered charges, for example water-supplies.

The environmental interest in this debate is confused. The 'polluter-pays principle' gives environmental credentials to road-pricing and other attempts to recover social-costs and externalities from private users. A scheme to add a levy on air-fares to pay for sound-insulation of homes under the flight-paths would be consistent with this thinking. However, if user-charges are applied at a flat rate, they become a regressive tax, like the British 'vehicle excise tax' of £110 per annum, per car, regardless of size, fuel consumption or the owner's income. Converting the flat-rate car-tax into additional duties on petrol would, at first sight, combine fiscal progressivity with environmental protection. By taxing gas-guzzlers, usually larger cars and/ or high performance cars, this measure would bear less hard on fuel-efficient and low-performance cars. On the other hand, such a scheme would tend to tax high-mileage rural drivers more than low-mileage town drivers, and would penalise essential users, for example people whose place or time of work precluded access by public transport.

THE CASE OF UNCLOS

The common heritage arguments deployed in the case of the seabed included the same combination of environmental management, equity and demilitarisation established as general characteristics above. In particular, the equity argument rested on recognising the historical advantages that the great maritime powers had gained, and any coastal state *might* gain from the existing law of the sea. International customary law was largely created by the major imperial, naval powers, principally in disputes between English and Dutch legal traditions during the seventeenth century. Customary law granted coastal states certain exclusive rights in the territorial sea, traditionally confined to three miles and, beyond, allowed all-comers free access to sea's resources. In practice, only a limited number of large coastal states possessed the naval and merchant marine power to exploit the free domains and, anyway, the trend of the post-1945 period was increasingly towards extending sovereign rights, potentially *ad infinitum*, in what has been previously referred to as creeping territoriality.

The Truman Declaration of 1945 extended US claims over the continental shelf. South American claims to 200-mile zones were made during the 1970s. The North Sea states annexed the seabed in the 1950s for oil-exploration, and Iceland made unilateral extensions of fishing-rights in the 1970s creating conflict with British claims to historic fishing-rights. These were all examples of the territorial trend. They were also instrumental in triggering confrontations, as opposite and adjacent countries, and states with closed or concave coastlines were confronted with the need to partition and demarcate waters that were previously international waters. The UK and Iceland, the UK and Argentina, the USA and Canada, Greece and Turkey, were all in dispute over maritime frontiers.[9]

The diplomatic origin of the common heritage concept can be clearly attributed to Malta's UN Ambassador, Arvid Pardo, who addressed the General Assembly on the topic in 1967. The Assembly adopted Resolution 2749 (XXV) on 17 December 1971 by a vote of 108 in favour, 0 against and 14 abstentions. The resolution established clear language on the CHM principle.

> The seabed and ocean floor, and the sub-soil thereof, beyond the limits of national jurisdiction as well as the resources of the area are the common heritage of mankind. (Preamble)

Other clauses referred to the reservation of the area for exclusively peaceful purposes and that its resources be used 'for the benefit of mankind as a whole'. The resolution therefore addressed both the questions of legal title and distribution of wealth.

The common heritage argument sought to preserve some part of the resources of the seabed for landlocked states, a total of 27 countries at that time, and also sought rights for so-called 'geographically disadvantaged' states, namely those with concave coastlines, or disproportionately small coastlines compared to their land-area, such as Iraq, Germany, Zaire; and for those countries, newly-independent after several centuries of colonialism, which, by definition, had taken no part in drafting the existing body of customary law, with its strong cultural bias towards rights of occupation and exploitation. The exploitation of marine resources which was envisaged in the law of the sea negotiations included non-renewable resources of the area, such as manganese, cobalt and copper recoverable from polymetallic nodules on the seabed.[10] The Convention also sought to control rates of exploitation of certain renewable resources. Highly-migratory fish-stocks, and marine mammals were specifically cited for protection, on grounds of maintaining catches at 'maximum allowable', that is, sustainable levels. These obligations not only applied to the high seas, but to states' practice within their 200-mile exclusive economic zone, (EEZ).[11] The high seas account for only 5 per cent of the world's fish-catch.[12] Most shallow-water fisheries are encompassed within the coastal states' expanded reach allowed by the 200-mile EEZ. However, high seas fish and migratory fish stocks are especially vulnerable. Their breeding cycles, rates of replenishment and therefore limits of sustainable, or recoverable, catch are subject to doubt and dispute. The precautionary principle would suggest the need to err on the side of caution, that is, on the side of the fish. High-seas fish-stocks are by definition the least protected of all stocks, and may be exploited faster than their replacement-rate, to the detriment both of its consumers (the 'tragedy of the commons' scenario) and of the ecology of the commons. By establishing a regulatory system of licences, quotas and taxation for the recovery of seabed minerals, the UNCLOS approach sought to derive an economic rent from the commons. Therefore, both equity and environmental conservation were served by proposing to regulate resources which were previously considered *res nullius*.

The common heritage concept was not without detractors, and the extended negotiations were, after a decade, thrown into confusion when the newly-installed Reagan Administration insisted on a clause-by-clause review of US commitments, on taking office in January 1981. The Reagan Administration focused their criticism on provisions for international control of seabed mining in the common heritage area, beyond the limits of national jurisdiction.[13] Also, by extending their territorial claims to 12 miles, and the EEZ to 200 miles, coastal states conflicted, in a fixed-sum manner, with the treaties' claims to the common heritage. Every mile gained for national jurisdiction was a mile less *available* for the common heritage. In fact, the coastal states pursued their interests so vigorously that the Convention, as finally adopted, did more to secure the sovereign rights of coastal states than to secure the ocean floor for the common heritage. As well as enlarging and standardising territorial claims, and creating the wholly new concept of the Exclusive Economic Zone, other provisions allowed the coastal states to extend their jurisdiction over the continental shelf to an absolute limit of 350 miles.[14] Approximately one-third of the ocean was thus reserved for the exclusive use of coastal states, including all shallow, inshore and exploitable waters. The 'area beyond national jurisdiction' only started beyond 200 miles (and in some cases 350 miles). It was the biggest land-grab in history, greatly exceeding the partition of Africa by the Berlin Congress of 1888. The largest beneficiaries were, of course, the largest coastal states, and some, such as France and UK, acquired the EEZs of their dependent territories such as the Falklands, Bermuda and New Caledonia.

Table 3.1   Area of EEZ in million square nautical miles

| United States | 4.82* | USSR | 1.26 |
|---|---|---|---|
| France | 2.86 | Japan | 1.13 |
| Australia | 2.41 | Brazil | 0.92 |
| Indonesia | 1.57 | Mexico | 0.83 |
| New Zealand | 1.41 | Denmark | 0.71** |
| United Kingdom | 1.34 | Papua-NG | 0.69 |
| Canada | 1.29 | Chile | 0.66 |

*Includes Northern Marianas and Micronesian trust territories which became independent in 1991.
**Includes Greenland.
*Source:* C. Sanger, op. cit., p. 65.

Therefore, after a twelve-year process of negotiation in UNCLOS (1970–82), the Convention was opened for signature at Montego Bay, Jamaica. It contained provisions for a common heritage regime for the sea-bed beyond the limit of national jurisdiction. It established, in embryo, a system of UN-administered licensing and revenue-sharing for the anticipated flow of wealth that would be forthcoming from the area.[15] Several setbacks have left these provisions in a state of suspense. Briefly, the USA, UK, Turkey and Israel declined to sign the Convention. So, many other states, which had signed, declined to ratify it, so that in 1992, only 45 of the necessary 60 instruments of ratification had been received. Thirdly, the political compromise which secured the common heritage clauses required the concession by the coastal states of a 200-mile exclusive economic zone, which at a stroke reduced the area liable to the common heritage regime by one-third. Fourthly, the recession in commodity prices experienced throughout the world economy after 1980 reduced the potential profits from seabed recovery of precious metals well below those obtained more easily from land-based production.

It is easy to see how the whole scheme of the common heritage of mankind fell foul of the neo-Conservative critique of the Reagan–Thatcher period. To some, marketising the seabed was clearly more acceptable than creating an international regulatory framework.[16] In Britain, an administration dedicated to 'rolling back the frontiers of the state' viewed the proposals for the ISA in the same light as the EC Commission. In short,

'Why let socialism in through the back door while ejecting the same from the front?' It was an ironic sense of 'rolling back the state' that could in the same breath help itself, so conveniently, to a 200-mile exclusive economic zone. The CHM clauses to which the US administration objected had not been written in blood by Third Worlders, but had, ironically, reflected much patient diplomacy and compromise by Henry Kissinger, acting as Nixon's and then Ford's Secretary of State. It was not an American, nor even a Republican objection, it was a faction within the Republican Party, which maintained their opposition to the treaty throughout the 1980s.[17] William Safire described the UNCLOS process as 'history's greatest attempted rip-off' and denounced the proposed regime 'operated under the United Nations and dominated by the world's socialist nations'.[18] In fact, the provisions for the membership of the proposed International Seabed Authority (ISA) were unexceptional in terms of the geographical distribution of seats.[19] UNCLOS was a 'rip-off', but one made by the coastal states, which treated the Convention *à la carte*. They enacted domestic legislation to take possession of the 12 and 200 mile zones; they continued the annexation of the continental shelf as in the North Sea; and, by not ratifying the treaty, they ensured that the provisions for sharing the CHM mining revenues would never come into operation.

THE NEW COMMONS?

Although this account of the UNCLOS process might suggest a dampening of enthusiasm for the common heritage of mankind, the idea has resurfaced in the context of new environmental concerns. Other commons have been cited earlier in the cases of Antarctica and outer space. Similarly, the 1986 decision of the International Whaling Commission to declare a moratorium on commercial whaling served to strengthen the environmentalists' claim that whales are no longer *res nullius*, and free to be hunted. Rather they are a common-property resource, requiring conservation, creating obligations and constraints on the behaviour of sovereign states. The Vienna Convention, relating to the ozone layer, and the Framework Convention on Climate Change of 1992 have elevated the atmospheric system into a common-property resource. The Vienna Convention (1985), Montreal Protocol (1987), and London Amendments (1990), between them have established a regime for the ozone layer, which seeks to protect it from damage by the continued release of chloro-fluoro-carbons (CFCs).[20] In other words, rather than regarding the ozone layer as *res nullius*, a free resource, it has been recognised, like the oceans, as a life-support system. As a common-heritage territory it has gained protection as states are now bound by obligations towards it.

The regime is complex, and recognises the difficulties faced by developing countries in finding non-CFC substitutes for many basic industrial and domestic uses, taken for granted in the industrialised countries for thirty years. Refrigeration technology is a widely-cited case. China and India between them have a human population of two billions (36 per cent of the world's total), who are poised on the brink of mass domestic refrigeration. If pursued with established cheap CFC refrigerants, the result would undo the reduced emissions and provisions for phasing out CFC production and use in the industrialised world, contained in the Vienna/Montreal/London regime. A compensation fund has therefore been agreed to assist these countries. The fund, donated by the larger CFC users among the developing countries, and controlled by the World Bank, UNDP and UNEP, was set at an initial $160 million, and contained provisions to rise to $240 million, on condition that China and India acceded to the regime.[21]

The protection of the ozone-layer is also related to the question of technology transfer. The environmental imperative is to encourage use of the best available technology. The developing countries wish to overcome the patent-protection and copyright costs incurred in adopting new technologies. For their part the leading innovative corporations wish to preserve the rewards of innovation, and, as in the case of the pharmaceutical companies, the major corporations claim that the profits derived from copyright and patent-protection are a necessary inducement for research, and a source of revenue to cross-subsidise blind-alley projects. The Montreal Protocol made provision for the transfer of technology to developing countries on the most favourable conditions.

Related to the ozone question has been the process, rapidly accelerating since 1988, of negotiating an international agreement on climate change. The demilitarisation of the world's climate system had been addressed as early as 1977, in the Environmental Modification Treaty (ENMOD), which outlawed climate modification for military purposes.[22] Despite enormous scientific controversy, the work of the joint UNEP–WMO Intergovernmental Panel on Climate Change (IPPC), reported in September 1990 a consensus view that numerous human activities were likely to cause measurable and destabilising climatic change over the next century. The cooperative management of manmade threats to climate stability has therefore become another dimension of the common heritage of mankind. Attention was focused on the role of carbon dioxide ($CO_2$) emissions from the combustion of hydrocarbon fuels in automobiles, industrial and domestic consumption, and from forest-burning. Other greenhouse gases were identified as methane released from coalmining, rice-production and the world's enlarged ruminant population, also CFCs, which have powerful heat-trapping as well as ozone-depleting potential. The IPCC formed the basis of intense negotiation in the preparatory phases of the UNCED which culminated in a major eve-of-conference furore, as concessions were made to accommodate US objections to language on quantitative targets for emission reductions that had been previously agreed among the EC12. The Framework Convention on Climate Change, adopted at Rio in June 1992, was thus flawed, but committed the parties to continued negotiation. The issue will be discussed more fully in Chapter 5.

Perhaps ironically, during the decade in which the pioneering Convention on the Law of the Sea has not yet been ratified by a sufficient number of states to bring its provisions for the common heritage into effect, two closely-related atmospheric commons have been recognised and, in one case, namely actions to protect the ozone layer, has already involved the creation of a complex regime, twice subjected to comprehensive extensions and amendments.

## SOVEREIGNTY *VERSUS* THE COMMON HERITAGE?

Sovereignty over natural resources is a fundamental demand of all countries. It is a particularly sensitive question for developing countries. Permanent sovereignty over natural resources is a logical component of self-determination. The sense of vulnerability of the Third World countries stems from their recent and sometimes bloody struggle to acquire independence since 1947. Specific and explicit reaffirmation of the concept has been a feature of postwar legal codes. Chowdhury cites sources such as the UN General Assembly resolution 626 (VII) of 21 December 1952, and drafts of the Human Rights Commission Covenant on Economic, Social and Cultural Rights dating from 1955.[23] The cases described above have all concerned attempts to create new legal obligations with respect to territories previously considered *res nullius*. Is it possible to extend the same arguments (environmental conservation, the indivisibility of benefits, demilitarisation, etc.) to sovereign resources and territories?

New claims to apply common-heritage principles to some countries' sovereign territories and resources have been advanced, most notably in the case of international calls on Brazil to restrain the destruction of the Amazonian rainforest.[24] The US government has, under pressure from NGOs, used its influence in the World Bank and Inter-American Development Bank to restrict loans to Brazil for new road projects.[25] The

Brazilian forests are neither *res nullius*, nor are they any other countries' property. Does Brazil *either* have some particular obligation towards the international community, *or* is Brazil violating some standard or norm of environmental management? Does this, rather like the question of Iraq and its treatment of the Kurds after March 1991, make Brazil accountable to some international authority for its actions, on an issue previously considered to be an internal affair of the state? McCleary argues forcefully, that Brazil is under *no* such general obligation. If some specific duty attaches to Brazil, it is one that should be properly applied elsewhere, to richer countries first.

> If Brazil has to deforest Amazonia to meet the basic needs of its people and develop economically, then the international community is obliged to aid Brazilians with their development. If the international community wants to preserve the rainforest for the well-being of its people, then it is obliged to assist Brazil not only in financing development but in creating alternate development strategies that are compatible with environmental preservation. Brazil, for its part, has a general duty to aid humanity but not at the expense of making its people worse off.[26]

If a sacrifice of income is to be made, there are many people a lot better off than the Brazilian poor.

The rainforests *are* a substantial sink for carbon dioxide emissions which could, very conveniently for the industrialised countries, be set against their carbon dioxide emissions in some comprehensive inventory of global sources and sinks. The tropical rainforests are also a reservoir of unique biological diversity, which may be harvested as a renewable resource. Furthermore, its largely unknown complexity may yield genetic materials of importance to medical science. These are all true. They may also be true of other vastly under-utilised resources such as the Siberian *taiga* and the Pacific north-west coastal temperate forests. These possibilities have not encouraged the governments of the Russian Federation, USA and Canada to entertain common-heritage ideas with respect to Siberia, Oregon and British Columbia.

The extension of common-heritage argument to sovereign resources has provoked a hostile response on grounds of national sovereignty, not only from Brazil, but also Malaysia and other tropical timber exporting countries. The Malaysian Prime Minister made a similar argument to McCleary in respect of biological diversity in his address to the Rio Summit.

> The poor countries have been told to preserve their forests and other genetic resources on the off-chance that at some future date something is

discovered which might prove useful to humanity. This is the same as telling these poor countries that they must continue to be poor because their forests and other resources are more precious than themselves.[27]

In the case of Brazil, and the African countries cited in Chapter 2 (if not Malaysia), growth is also needed to meet the burdens of debt-repayment. Some countries that are most assertive on the question of sovereignty are among the most heavily indebted. The linkage may be direct, as in the rate of commercial timber-extraction associated with Mayanmar (Burma) and Thailand. Alternatively, the linkage to debt may be only indirect; in Brazil only a very small percentage of deforestation is attributable to commercial logging. The forest is the last free frontier for the landless, the dispossessed and the unemployed victims of Brazil's structural adjustment to its debt burdens. The resistance of the tropical timber exporting countries to being made scapegoats, and to any diminution of their sovereign rights, was particularly evident at Rio. Throughout the preparatory process, especially after Geneva in August 1991, it was apparent that the Convention on Tropical Forestry as proposed by the conference secretariat would be unacceptable. The eventual compromise at UNCED produced a *non*-legally binding Statement of Principles, applicable to temperate *as well as* tropical countries. This will be further explored in Chapter 5.

The new-found Northern enthusiasm for the preservation of the Third World's environmental quality, may be perceived as hypocritical by the subjects of this new eco-colonialism. The North, having spent the past 400 years going about the destruction of their own temperate forests, and 200 years in the production of carbon dioxide emissions since the birth of the industrial revolution, now desire to protect other people's forests and furry animals for the so-called common heritage of mankind. Where, may ask the poor of Brazil and Africa, is the UK commitment to preserve North Sea oil for the common heritage of mankind? When burnt it will be a potent source of carbon dioxide emissions. It, like the rainforest, has a host of non-fuel uses, as a raw material for plastics and fertilisers. North Sea oil is also a precious natural resource that one country reserves the right to exhaust over a limited time-frame regardless of the poor countries' interest in the redistribution of income, and also regardless of the trans-generational interests of Briton's yet unborn; that is, a slower rate of exploitation of North Sea reserves would allow the benefits to accrue to a future generation and so be justified on grounds of sustainability and stewardship. In plain English the depletion of a non-renewable resource is not sustainable indefinitely. Such depletion can only be reconciled with the looser requirements of 'sustainable development' in two ways. Either the depletion of fossil fuels is phased against the development and introduction of renewable energy sources with an equivalent calorific yield; or, alternatively, the rate of depletion is slowed, so as to benefit at least more than two generations, however defined.

The suggestion that some other country's natural resources may be claimed for the common heritage is especially controversial following the decision of the USA, UK, the other G7 powers and all of the EC12 *not* to ratify the 1982 Convention on the Law of the Sea. These countries have undermined common-heritage arguments by their selective utilisation of the convention's provisions for the *extension* of sovereign rights over territorial waters, the 200-mile Exclusive Economic Zone, and maximum possible 350-mile claims to the Continental Shelf. An attempt by a small number of equatorial Third World countries to apply the same logic to outer space, in the claims of the Bogota Declaration to extend air-space into geostationary orbits space, has been unsuccessful.[28] All other things being equal, extending territorial rights need not be detrimental to environmental conservation. An enforceable EEZ is a better protection for fish stocks than a spurious international administration that is unenforced. However, it does contradict the rationale of arguing that other countries' forests and other countries' biological diversity are somehow candidates for the common heritage. The environment may be well served by the extension of national sovereign rights. Tough regulatory frameworks, such as those associated with Scandinavian countries, Canada and the EC in some respects, can, like the IWC case, use domestic law and regional agreements such as the Paris Conventions on the North Sea to protect the commons. The problems of distributive justice remain unaddressed, as does the practical impossibility of nationalising or indeed 'privatising' the weather, or outer space.

## A COMPROMISE?

All that said, it is also true it *would* be better, all other things being equal, if the forests of the Amazon were saved rather than depleted. The rainforests of Brazil *could* be protected in a way which might satisfy the interests of both Brazil and the industrial powers. The latter could *pay* Brazil a reasonable compensation *not* to cut down the trees. Debt-for-nature swaps are one way in which this might be happen. The American and European governments and NGOs could purchase

Brazilian debts from the private banks, and write them off, in return for agreed and verifiable conservation commitments. Debt-for-equity would allow the politically more disturbing option of outright purchase of the forest-regions, acre-by-acre, to leave them unexploited. A genuinely market-led option would be for the government of Brazil to sell tracts of forests for European countries to purchase, and conserve, to claim them as a carbon-sink, or 'bank' to set against their industrial and automobile emissions, in some future climate-change treaty. If tradeable permits for emissions are widely canvassed, why not tradeable permits for sinks?[29] At the NGO level, outright purchase for conservation has been canvassed by a British charity seeking £1.25 million for the purchase of 100 000 acres of tropical rainforests in Belize.[30]

Another basis for *negotiation* between the developing countries and the developed *may* exist to the extent that the former regard environmentally sound technologies as falling into the domain of the common heritage of mankind. More specifically, *access* to the best available technologies on concessional terms, i.e. technology-transfer invoking common heritage arguments, would *allow* many developing countries to apply environmentally responsible solutions that would otherwise be beyond their means. The developed countries regard new technologies as private property, subject to copyright and patent protection, i.e. intellectual property. Non-CFC substitutes, fluidised bed-combustion equipment for efficient coal-burning, and catalytic converters for motor vehicles, are all examples of technologies which western countries would *like* developing countries to adopt.

A willingness on the part of OECD countries to regard best-available technologies as suitable for transfer on preferential terms might be traded with the agreement of developing countries to accept common-heritage arguments in respect of their sovereign resources. Of itself this is not a new idea. Principle 21 adopted at Stockholm in 1972 contained the commitment that 'environmental technologies should be made available to developing countries on terms which would encourage their wide dissemination without constituting an economic burden on the developing countries'.[31] This issue particularly alienated the Reagan Administration from signing the Convention on the Law of the Sea in 1982. Its provisions for UN-licensed deep-sea mining contained provisions for private consortia to transfer technology to the ISA. A decade later, the Rio Declaration made no specific reference to technology transfer. However, the *Agenda 21* devoted a full chapter to the issue, including the commitment to 'favourable access to and transfer of environmentally sound technologies, in particular to developing countries'.[32] The grounds for negotiation and compromise may be represented as in Table 3.2.

Table 3.2   The growth of the common heritage of mankind

| *Res nullius* | *common heritage of mankind* | *national sovereignty* |
|---|---|---|
| The high seas | Deep ocean floor | Territorial seas and EEZs |
| Whales, pre-IWC moratorium | Whales after moratorium | |
| Highly-migratory species pre-1982 | Highly-migratory species under UNCLOS terms | |
| Outer space pre-1967 | 1967 treaty provisions | |
| Ozone layer pre-1985 | Montreal protocol terms | |
| Climate system now | Possible $CO_2$ limits negotiated in FCCC? | |
| | 1979 radio-frequencies allocation by WARC | Jamming, conflicting national use of wavelengths. Bogota Convention claims to GSO. |
| Unclaimed parts of Antarctica | 1959 treaty provisions | Parts of Antarctica still claimed |
| | Technology transfers? | Copyright and patents (intellectual property) |
| | Possible tropical forestry? (world-heritage sites) | Present tropical forestry (national resources) |

WARC = World Administrative Radio Conference. GSO = geo-stationary orbit.

In this scheme, environmental diplomacy on the ozone layer and climate-change constitute attempts to shift these commons from *res nullius* status to that of the common heritage. More controversial are arguments to shift presently *sovereign* resources, such as Brazilian rainforests, into the category of the common heritage on some principle of stewardship. Without financial compensation, or a trade-off involving access to desirable artifacts of intellectual property, such proposals appear unworkable.

Title to property carries the right to retain the revenue derived from its use. Private property generates private revenues, although these may be subject to taxation. Logically, common-heritage property yields income for the international community which has title to that property. *Res nullius* generates no income, it simply *allows* others to draw upon it. Accepting the principle of user charges for the commons could generate the resources necessary for sustainable development. As discussed earlier, the revenue-sharing aspects of the regime proposed for the seabed were among the obstacles to securing American and British endorsement of the 1982 Law of the Sea convention.

Ironically, the British government came later *to endorse* the view that environmental capital must be priced, and the 'polluter-pays-principle' has, in name at least, penetrated the department of the Environment. The Pearce report has shown free-marketeers that the market has limits, and that a proper market valuation of previously free resources may be consistent with conservative principles. This suggests that it is *politically* propitious to reconsider the wider application of the principle of the common heritage of mankind.

A number of ambitious proposals for deriving and applying revenues from the commons will be discussed in the Chapter 6. These may be linked to the question of debt-relief. These revenues could be applied to research and action to extend the conservation of the commons, for example, in ozone-layer and related climate research. Alternatively, revenues from international carbon taxes, etc. could be directed to a sinking fund for relief of Third World debt. Finally, such revenues might provide funds for a massive expansion of UNEP and UNDP activities towards implementing sustainable development. Before these opportunities can be discussed, the path of UN efforts in the field from Stockholm to Rio must be addressed and considered against the imperatives identified in this and the preceding chapters.

# 4 The UNEP Role

The labours of Hercules were essentially heroic. Charged with drama, they hold the attention of the reader and end happily. Sisyphus, condemned for eternity to roll a rock to the top of a cliff, only to have it roll back, so that he had to start again, commands our attention as a metaphor for futility. Put simply, this chapter argues that the United Nations Environment Programme's so-called catalytic mandate was a Herculean challenge, whereas its so-called coordinating mandate, contained in the same 1972 founding resolution, was Sisyphean. To perform the first, UNEP must be released from the nightmare of the last. Some progress in this direction was conceded at Rio by the decision to create a new Commission on Sustainable Development (CSD), that will free UNEP for more creative tasks, but also expose it to new risks, to be fully explored in Chapter 5.

## THE STOCKHOLM PRINCIPLES

The creation of the United Nations Environmental Programme was a direct consequence of the 1972 United Nations Conference on the Human Environment, held at Stockholm, 5–16 June 1972. McCormick emphasises the role of a single issue, acid-deposition or 'acid rain', in concentrating minds on the need for such a conference. He also cites the apprehension of the developing countries that an agenda led by pollution issues, would marginalise the poor countries.[1] The parallel with preparations for Rio 1992 is instructive. The preoccupation of the western countries to tackle the question of climate change was not the most pressing concern of the Third World. The western desire for a narrowly environmental conference was linked by Third World governments to the wider development agenda.

Stockholm, like Rio, benefited from a massive preparatory effort. In addition to formal sessions of the preparatory committee in Geneva and New York, subsidiary expert meetings such as those at Founex, Switzerland in June 1971 and Canberra in August–September 1971, contributed to the agenda and planning. Other caucuses, such as the G77 meeting at Lima, Peru, in November 1971 stressed the importance of Third World interests. The Preparatory Committee was limited to 27, whereas UNCED elected 39 Vice-Chairmen alone. McCormick credits Maurice Strong with accepting and incorporating the Third World views, to the extent that the agenda of the conference and the very concept of environment were broadened to include issues such as soil loss, desertification, tropical eco-system management, water supply and human settlements.[2]

The Stockholm preparatory process was not without explicit political disagreement. In December 1971 the United States and the United Kingdom both voted against a UN General Assembly resolution commending the work of the preparatory committee and the conference mandate. Resolution 2849 (XXVI) specifically called for prohibition of testing and production of nuclear weapons. The resolution was adopted by a vote of 85 to 2, with 34 abstentions. The Soviet Union and their East European allies also suffered a disappointment. The General Assembly adopted Resolution 2850 (XXVI), concerning the draft agenda and rules of procedure. This resolution restricted participation to UN member states, and the members of the specialised agencies. Whereas neither of the German states was, at that time, a member of the UN, West Germany was a member of many specialised agencies, and so eligible to attend, whereas the GDR was not. A Soviet bloc amendment which tried to postpone the conference until 1973 was defeated. The original resolution was adopted by 104 to 9, with 7 abstentions.[3] The Conference itself witnessed many intense political attacks on the behaviour of some participants. The US was attacked for its conduct of the Vietnam war, while China repeatedly attempted to insert the language of neo-colonialism.

Concerning UNEP, U Thant, as early as 1971, spoke of his preference for creating a 'switch-board' to link the existing agencies, rather than creating a 'super-agency'. McCormick cites Brian Johnson's contemporary suggestion that the other specialised agencies within the UN system actively discouraged the creation of any such agency. In fact, as it emerged, UNEP closely resembled American preferences for a programme supported by net additional funds on a voluntary basis, linked to a small but powerful executive within the office of the Secretary General.[4] The Final Declaration of the Stockholm Conference constituted a call to action. Specifically it cited 109 recommendations and the 26 Principles. The latter comprised a fascinating combination of aspiration and obfuscation relating to the protection of the environment. Certain points are of particular interest to compare with the development of UNEP's capacity in the years after 1972. China was the only participant to withhold its support from the final vote on the Conference Declaration.

The 26 principles can be grouped in terms of their political objectives and content. This may be abbreviated to S1 etc, for Stockholm Principle 1, etc.[5]

*Political Declarations: S1, S26*

The first of 26 principles contained an explicit political statement linking human rights, 'dignity and well-being', to the 'solemn responsibility to protect and improve the environment for present and future generations'. It continued, rather grandly,

> In this respect policies promoting or perpetuating *apartheid*, racial segregation, discrimination, colonial and other forms of oppression and foreign domination stand condemned and must be eliminated.

Principle 26 made a similarly-styled denunciation, that 'man and environment must be spared the effects of nuclear weapons', and concluded, 'states must strive to reach prompt agreement, in the relevant international organs, on the elimination and destruction of such weapons'.

In between the first and the last, the other principles wove together an enormous variety of definitions and commitments more obviously within the scope of the conference mandate.

*The Commitments to Conservation: S2, 3, 4, 5, 6, 7*

The second principle showed the extent to which the proceedings had focused, as the title suggested, on the *human* environment. The commitment to preserve 'especially representative samples of natural ecosystems' was for the 'benefit of future generations'. It has taken two decades of animal rights activism and green political thought to establish the distinction between environmentalism and ecologism noted by Dobson in Chapter 1. Perhaps also representative of the times, S2 and S4 separated the survival of such 'representative samples' from the question of their habitat. Both were noted, but as separate issues, as if parrots could live without forests. The linkages inherent in the concept of biological diversity, combining habitat, the survival of species and limitations on the trade in endangered species, as applied to the ivory trade, and spotted cat skins, has taken many years to develop.

S3 and S5 made the distinction between renewable and non-renewable resources. The statement on renewables was self-evidential and uncontroversial. 'The capacity of the earth to produce vital renewable resources must be maintained and, wherever practicable, restored or improved.' However, the statement on non-renewables, S5, hinted at a particular view of property.

The non-renewable resources of the earth must be employed in such a way as to guard against the danger of their future exhaustion and to ensure that benefits from such employment are shared by all mankind.

Fifteen years in advance of the Brundtland report's notions of sustainable development, the implications of this pledge are in fact staggering. Was coal to be carted to Jakarta, or oil piped to Paraguay, to be 'shared by all mankind'? From each according to his ability, to each according to his need, may have had both Biblical and Marxist belief on its side, but was contrary to statements elsewhere concerning the primacy of states' sovereign rights over natural resources. In S7, marine pollution received specific mention, and a tentative link was made to the question of territories *res nullius* in S21.

### The Rights of Developing Countries: S8, 9, 10, 11, 12, 20, 23

The explicit connection between environmental quality and development was reflected in a group of seven related principles. In a direct challenge to the no-growth advocates, S9 identified 'accelerated development' as the right and expectation of the developing countries. A commitment to the redistribution of 'substantial quantities of financial and technological assistance as a supplement to the domestic efforts of the developing countries' was included. This was linked, in S10, to commitment to support 'adequate earnings for primary commodities and raw materials'. The Stockholm Declaration was adopted just 15 months before the October 1973 initiatives by OPEC to raise the price of crude oil by 400 per cent. S23 further protected the LDCs from the financial burdens of regulation in its suggestion that

> it will be essential in all cases to consider the system of values prevail-
> ing in each country and the extent of the applicability of standards which
> are valid for the most advanced countries but which may be inappropri-
> ate and of unwarranted social cost for the developing countries.

### Planning and Government: S13, 14, 15, 17

S14 called for the 'integrated and coordinated approach' to development, and S17 for 'appropriate national institutions'. These were translated into one of the most widely-credited achievements of the Stockholm Conference. Many governments reformed the division of ministerial responsibilities so as to take specific account of the needs for integrated environmental planning. A widely cited statistic is that in 1972 only 25 UN members had environment ministries, but that by 1990 the figure was over 125.[6]

## Population: S16

Population growth was as controversial in 1972 as in 1992.[7] In the interim, world population rose from approximately 4 billion to 5.5 billion. Ninety per cent of that increase occurred in the developing countries. The Stockholm Declaration, Principle 16, was a very conservative statement:

Demographic policies which are without prejudice to basic human rights and which are deemed appropriate by governments concerned should be applied in those regions where the rate of population growth or excessive population concentrations are likely to have adverse effects on the human environment and impede development.

The *laissez-faire* of most governments and the draconian one-child policy of China both fall within the definition of policies 'deemed appropriate by governments concerned'. During the late 1980s the Reagan administration withheld funds from the UN Fund for Population Activities (UNFPA), because China used UNFPA monies for abortions.

## Property: S5, 7, 20, 21, 22

McCormick, in his summary of the Stockholm principles, refers to a commitment that 'non-renewable sources should be shared'.[8] This sentiment noted in Principle 5, was not developed beyond the form of a platitudinous statement. Specific opportunities elsewhere to build upon this concept contain mixed evidence for advocates of the common-heritage approach . S7 explicitly referred to the obligation to prevent marine pollution, which created 'hazards to human health, to harm living resources and marine life , to damage amenities or to interfere with *other legitimate uses of the seas* (emphases added). In 1971, at the very earliest stages of negotiations in UNCLOS, these 'other legitimate purposes' were an essentially *res nullius* attitude to fishing and other unregulated uses of the high seas.

The two most widely-cited of the Stockholm Declaration's 26 principles contain tentative moves towards the principles of common heritage, while simultaneously affirming the principles of sovereignty over natural resources. This can be seen in the provisions of Principles 21 and 22.

States have, in accordance with the Charter of the United Nations and the principles of international law, the sovereign right to exploit their own resources pursuant to their own environmental policies, and the responsibility to ensure that activities within their jurisdiction or control do not cause damage to the environment of other states or of areas beyond the limits of national jurisdiction.

The commitment to refrain from trans-boundary pollution, is consistent with the trend of international customary law since the Trail smelter case.[9] It was reinforced in Principle 22 which urged the signatories to

cooperate to develop further the international law regarding liability and compensation for the victims of pollution and other environmental damage caused by activities within the jurisdiction or control of such states to areas beyond their jurisdiction.

Thus sovereignty and international obligations were reconciled to the extent that the transboundary effects of environmental pollution are assigned to the responsibility of the polluting state. However, it would require a much stronger case to argue, for example, that destruction of the rainforests, however regrettable, constituted 'damage to the environment of other states'. Nothing here requires, say Brazil, to regard its forests as a common-heritage property, any more than, say the UK could be held to account for the now nearly-cleared Forest of Caledon.

As cited in Chapter 3, a more explicit treatment was given to technology transfer. S20 referred not only to supporting the 'free flow of up to date scientific information and transfer of experience' but also stated:

environmental technology should be made available to developing countries on terms which would encourage their wide dissemination without constituting an economic burden on the developing countries.

In practice this was a plea to trade intellectual property, that is, patent and copyright protection at less than its full market value. Since it was unrealistic to suggest that major corporations could operate a dual pricing policy, it implied that governments, or some international authority such as the UN, would 'buy-out' such intellectual property and incorporate its subsidised transfer into aid programmes. In the 1970s, widespread criticism of transnational corporations was in vogue, and focused upon their comparative freedom from regulation through transfer-pricing, banking secrecy laws in certain countries and the use of off-shore tax-havens. Supra-national control of intellectual property therefore fits with the idea of taxing these essentially global corporations through a global system, the UN, an idea which waned during the following decade of Reagan, recession and rescheduling.

*International Organisations: S24, 25*

The Stockholm Principles made specific reference to the foundation of the appropriate international organisations to carry forward the commitments undertaken by the members.

Co-operation through multilateral or bilateral arrangements or other appropriate means is essential to effectively control, prevent, reduce and eliminate adverse environmental effects resulting from activities conducted in all spheres, in such a way that due account is taken of the sovereignty and interests of all states. (S24)

States shall ensure that international organizations play a coordinated, efficient and dynamic role for the protection and improvement of the environment. (S25)

The Stockholm principles therefore recognised the bias towards the poor, and the obligation of states to respect the environmental quality of both their neighbours' territory as well as the commons beyond national jurisdiction, while simultaneously affirming the sovereign rights of states to control of their natural resources. The principles also recognised the inherent limitations of national actions on environmental questions and so endorsed the creation of appropriate, although very limited, UN mechanisms.

## UNEP's CATALYTIC ROLE

UNEP was created, as 'a small secretariat', as a direct consequence of the Stockholm Conference.[10] The UN General Assembly resolution established UNEP as an integral part of the United Nations, whatever impression its separate title and Nairobi location might suggest. UNEP's Executive Director is elected by the General Assembly, on the nomination of the UN Secretary-General; her salary and those of a number of senior UNEP posts are funded from the UN-assessed budget; UNEP reports to the General Assembly through ECOSOC. UNEP was expressly mandated to coordinate the activities of other UN organs, although whether the mandate was reasonable, or even credible, is subject to extended discussion later. The catalytic role is less explicit, indeed the term does not occur in the founding resolution, but has nonetheless acquired a central status. UNEP was not intended to operate as an *executive* programme, undertaking major scientific research in its own name. Rather it was created to act as clearing-house for environmental data and research and to operate, at most, demonstration projects which would establish the feasibility of schemes to be adopted, on a larger scale, by the member states and other UN organs. Typical of this division of labour is the separate existence of UNESCO's massive 'Man and the Biosphere Programme'.

UNEP is peculiar in having no Statute, Charter or Convention to describe its functions and parts. That role has been furnished for over twenty years by General Assembly resolution 2997 (XXVII), a lengthy, detailed statement of the Stockholm conference participants on the subject, adopted in the General Assembly on 15 December 1972. The resolution set out, in the manner of a statute, the structure and functions of the UNEP Governing Council, Secretariat and Fund. A lengthy preamble defined the ambitions of the members in the field of environmental diplomacy. Significantly, the members recognised the limit of *international* action, in observing that

> responsibility for action to protect and enhance the environment rests primarily with governments and in the first instance, can be exercised more effectively at the national and regional levels.

A second early recognition of UNEP's likely limitations was contained in the passage that referred to 'respect for the sovereign rights of states' and being 'mindful of the sectoral responsibilities of the organizations of the United Nations system'. Resolution 2997 also recognised the necessary connection between environment and development, and linked these to the proposition that 'in order to be effective international cooperation in the field of the environment requires additional financial and technical resources'. Attempts by the US, in particular, to promote the 1992 agenda on the basis of no net additional resources, were thus contrary to policies adopted twenty years previously. The preamble concluded by describing UNEP's rationale, recognising,

> the urgent need for a permanent institutional arrangement within the UN system for the protection and improvement of the environment.

UNEP is managed by a 58-member Governing Council elected from the General Assembly on a regional formula to serve a three-year term. This uses the simple five-regions system which exists within the UN, something older, and more ingrained in UN procedure than the so-called caucuses or party-politics groupings such as the Neutral and Nonaligned States (NNA), the Organisation of the Islamic Conference (OIC) and indeed the European Community (EC12). The regional formula gives 16 seats to Africa, 13 to Asia, 6 to Eastern Europe, 10 to Latin America and 13 for the ubiquitous West European and Other States Group (WEOG).

The Governing Council of 58 meets only biennially, although since shifting to this sequence it has met in Special Sessions in the years in between. Besides conducting the business of UNEP through meetings of the Governing Council, the Council has also created a quarterly series of meetings – the Committee of Permanent Representatives, or CPR, comprising the Nairobi-based missions. A 'Bureau' is formed for a biennium with a member of each of the five regional groups serving in the five positions of President, three Vice Presidents and Rapporteur.

The Governing Council is mandated to

promote international cooperation in the field of the environment and to recommend, as appropriate policies to this end; To provide general policy guidance for the direction and co-ordination of environmental programmes within the UN system.

To keep under review the world environmental situation in order to ensure that emerging environmental problems of wide international significance receive appropriate and adequate consideration within the UN system.

It is worth considering the implications of this passage for coordination within the UN system. This resolution, adopted by governments, created an organ of inter-governmental cooperation. The resolution mandates those governments to coordinate their own actions in the specialised agencies, such as the IAEA, and in other UN organs, such as the UN Development Programme (UNDP). It is a resolution in which governments pledge themselves to know the difference between the right hand and left hand, and furthermore to coordinate the actions of both. UNEP, in the sense of a corporate body, imagined as some autonomous agent, cannot necessarily do this. The member governments, sitting as members in different places such as the General Assembly, the FAO or the UNDP, are pledging *themselves* to do this. The scope for the Secretariat to act autonomously will be discussed shortly.

The Governing Council was also mandated to receive and review the reports of the Executive Director, and to

keep under review the world environmental situation in order to ensure that emerging environmental problems of world wide significance receive appropriate consideration by Governments (.)

This function, essentially one of early warning, is then linked to the role of the Governing Council in promoting

the contribution of the relevant international scientific and other professional communities to the acquisition, assessment and exchange of environmental knowledge and information (.)

Resolution 2997 made further references to formulating and implementing environmental programmes within the UN system, and to maintaining a 'continuing review' of the impact of national and international policies on the developing countries, and their plans and priorities.[11]

As mentioned earlier, although resolution 2997 made numerous references to coordination, it made no use of the term 'catalytic', which emerged subsequently as a synonym for the mandate to promote and encourage the attention of membership on environmental questions. The catalytic role is *implied* in Section I, 2, b, which referred to the *direction* and coordination of environmental programmes (emphases added) and II, 1, which described the Secretariat as a 'focal point for action in the United Nations'. The expression is now well-established, as in the General Assembly resolutions of 1987–88, namely 42/184 and 42/186 of 11 December 1987, also 43/196 which mandated the UNCED, and describes UNEP in the Preamble as having a 'catalytic role in technical cooperation among developing countries in the field of environment'.

The UNEP Secretariat is mandated, in Resolution 2997, to

> serve as a focal point for environmental action and co-ordination within the United Nations system in such a way as to ensure a high degree of effective management (.)

The Executive Director (ED), is specifically entrusted with a number of responsibilities. The first five of these may be characterised as executive (in the British sense), placing the Executive Director in the role of servicing the Governing Council and implementing its mandate established earlier. The ED's role is to 'provide substantive support to the Governing Council', to coordinate programmes 'under the guidance of' the Council, to offer advice to other UN organs, to secure the cooperation of the international scientific community, and to provide, on request 'advisory services for the promotion of international cooperation in the field of the environment'.[12] Thereafter, the ED's responsibilities become more executive in the American sense, that is discretionary, creative and even innovative. Two key paragraphs define this role. The ED is

> To bring to the attention of the Governing Council, on his own initiative or upon request proposals embodying medium-range and long range planning for the United Nations in the field of the environment;

> To bring to the attention of the Governing Council any matter which he deems to require consideration by it (.)[13]

This is an invitation for a dynamic and ambitious ED to largely define the job as she or he thinks fit. Taken together, the description of the ED's responsibilities combines elements of two models of UN practice, the bureaucratic and the charismatic. The original distinction concerned British and French styles in the international civil service, as they developed in League of Nations practice after 1919. Drummond, of the League, was characterised as the quintessential British civil-servant, the adviser, dispassionate, both very civil and very much the servant. This contrasted with the model of Albert Thomas, an early Director-General of the ILO, who reflected the preference of the French civil service to be politically committed, and explicitly guiding policy. Within the limitations of these stereotypes, Mostapha Tolba, the UNEP Executive Director, who dominated the organisation during 1976–92, clearly fits the charismatic model.

Dr Mostapha Tolba succeeded the first ED, Maurice Strong, who had served as Chairman of the Stockholm Conference and was to return twenty years later in the same role at Rio. In his turn, Tolba was succeeded by Dr Elizabeth Dowdswell, a Canadian, on 1 January 1993. Tolba is an Egyptian, currently serving as his country's Ambassador to the UN, who was born in 1922. He was appointed as Executive Director on 1 January 1977 and served four four-year terms. Tolba received his first degree at Cairo, and his PhD from Imperial College, London. He held academic posts in Cairo and Baghdad before joining the Egyptian government and diplomatic service. Numerous individuals have attested to the strong personal style of Mostapha Tolba's term of office, to his extraordinary work-rate, his reluctance to delegate authority, in particular in the matter of appointments, and to his single-mindedness in the promotion of the work and role of UNEP. Both praise and criticism were applied to his reluctance to delegate authority.

It is always instructive to follow the flow of money; it is the test of priorities in a political system where rhetorical claims are free and hence frequent, but where spending money is painful and so occurs in a two-yearly cycle. The Programme Budget for 1990–91, the last full cycle before the onset of UNCED responsibilities, set a spending target of just $68 million. The Environment Fund is itself raised by voluntary contributions by the members.

Table 4.1    The leading contributors to the UN Environment Fund, 1987
(percentages)

| US | 21.9 | France | 3.8 |
|---|---|---|---|
| Japan | 14.5 | Norway | 3.4 |
| USSR | 13.2 | Canada | 2.6 |
| Sweden | 8.5 | Netherlands | 2.6 |
| W. Germany | 8.4 | All others | 16.0 |
| UK | 5.1 | | |

Table 4.2   The UNEP budget

| Programme | Share of budget as % |
|---|---|
| 1. Atmosphere | 4 |
| 2. Water | 5 |
| 3. Terrestrial ecosystems | 17.5 |
| *in which* desertification | 9.1 |
| 4. Oceans | 10.3 |
| 5. Lithosphere | 0.9 |
| 6. Human settlements | 1.9 |
| 7. Human health/welfare | 2.2 |
| 8. Energy, industry and transport | 6.3 |
| 9. Peace and security | 0.6 |
| 10. Environmental assessment | 21.8 |
| 11. Environmental management | 5.7 |
| 12. Environmental awareness | 13.6 |
| 13. Technical and regional cooperation | 10.6 |

By examining the Programme Budget a detailed picture of UNEP priorities can be discerned (see Table 4.2).[14] Three characteristics are apparent. First $68 million is a very small sum of money, compared to national budgets and corporate spending. Second, it is noticeable that a high proportion of UNEP funds are targeted on a small number of programmes. In annual, sterling terms the largest item, Programme 10, was budgeted at approximately £4.25 million. In the USA, the Global Change Research Programme, requested by the President in his FY 1991 Budget requested Congress to fund the programme at $1034 million, an increase of $344.8 million or 57 per cent on the previous year, in a period of budgetary austerity. (UNEP's annualised Programme Budget, of $34 million for that year, was therefore just 3.28 per cent of the US government's $1034 million *national* research budget for cognate activities.)[15] Third, Programme 10, including INFOTERRA, IRPTC, and GEMS/GRID, took over one-fifth of the programme budget. The terrestrial ecosystems, Programme 3, took just under one-fifth of the programme budget. The oceans, Programme 4, took one-tenth. The combined share of these three programmes was very nearly half of the total programme budget (49.6 per cent).

Programme 10 fulfilled UNEP's obligation to become a centre of data-gathering on the state of the environment, a 'one-stop shopping centre' for basic scientific knowledge. The concentration on arid lands is a clear priority: the draft budget set an objective to provide assistance to 15 countries affected by soil degradation and desertification. The total sum involved was just $7.2 million. Also contained in Programme 3 was the provision of just $1.05 million to sub-programme 3.3 for tropical forest and woodland ecosystems.

Smaller science elements such as Programme 1 on the atmosphere and Programme 5 (on the lithosphere) were therefore smaller than the public information budget, set at $5 million, as sub-programme 12. 2. This is not peripheral, but central to the UNEP's mandate to communicate the results of its projects, not just to the UN and the member states, but to the NGO and wider public as well. Given that the sums involved are, in absolute terms, the small change of OECD-member budgets it is not surprising that, during the period 1990–92, Dr Tolba made repeated calls to raise the Environment Fund to $250 million within a decade. However that raises the question, addressed in Chapter 5, that an expansion of UNEP's catalytic role might lead to replication of the work of the larger specialised agencies and the UNDP.


EVALUATING UNEP

UNEP has not escaped criticism for the *proportion* of its funds which are expended on programme and programme support costs or PPSC. This is an issue as old as the UN itself and was used, somewhat speciously, in the Reagan Administration's attack upon UNESCO. PPSC represents that part of the budget allocated to the secretariat's operating costs (except that part which is contributed by the UN-assessed budget). In the biennium 1988–89, the Environment Fund was set at a total of $87.8 million, of which PPSC was set at $25.8, the Programme itself was allocated $60 million, plus a reserve of $2 million, and finally a financial reserve, set at 7.5 per cent or $6.6 million. Of the PPSC's $25.8 million, $17.5 or 68 per cent was allocated for salaries, and the remainder for travel, equipment, supplies and all other operating costs (rent, utilities, posts and telecommunications, etc.). UNEP is mandated by the Governing Council to maintain PPSC at no more than 33 per cent of contributions to the Environment Fund. The estimate of the United States General Accounting Office was that on the basis of an anticipated $60 million contribution, PPSC had risen to 43 per cent of contributions. At the time, 1989, particular attention was focused upon liaison and regional offices as likely candidates for cost-cutting and closer financial scrutiny.[16]

Another line of criticism that has been advanced concerns the trade-off between breadth and depth in UNEP's activities. The GAO reported the unattributed comments of US government and NGO officials that UNEP is in too many areas and is oversubscribed, its resources are spread too thin, and its funds farmed out to too many UN agencies, making it difficult for the organisation to be really efficient.

The overall conclusion gave credit to UNEP, as being 'generally successful at carrying out its catalytic and coordinating functions' but remarked,

> UNEP, however may be dissipating some of its resources by attempting to administer too many small-scale projects which have marginal impact.[17]

The observation that the Programme is, in essence, spread very broadly but of very restricted depth, is an almost inevitable criticism given the global mandate to conduct investigations across the whole range of environmental sciences, and from the atmosphere to the lithosphere, but to do so with selected pilot projects testing and demonstrating the application of techniques.

Something of the flavour of UNEP's work and its operational achievements and failures can be gleaned from the following comparison of two superficially similar projects. Both concerned forestry development in Asia and demonstrate the fine margin between earning praise or criticism. An example of a well-received project which illustrates the catalytic mandate, and the very small sums of money involved was a 1985–87 demonstration project on community afforestation and training in southern India. UNEP contributed just $50 817 to a $71 000 project. Over a period of two years, in addition to the original target of 150 farmers, 575 other farmers were trained in tree-planting and propagation, 1000 school children were trained to create tree seed-beds, four tree nurseries managed by local women were established, and 42 'people's nurseries' were established. Finally, 750 000 trees were planted, and a documentary film was made for local village viewing.[18]

Not all projects are so successful. A superficially similar afforestation project in Yemen was described thus by UNEP's own Evaluation Report: 'taking account of the time and funds involved the outputs produced cannot be considered cost-effective'.[19] The project, in the Lodar district of Yemen, expended $1.9 million (UNEP's contribution being 49 per cent), over five-and-a-half years in a project to establish an agro-forestry complex on a 390-hectare area of poor, dry soils, and to test different techniques of cropping, tillage, fertiliser-application and irrigation under local conditions. The Evaluation Report detailed overruns in the timing of the project from 24 months to 68 months, delays in recruiting foreign experts, delays in the supply of heavy earth-moving equipment, the frequent mechanical breakdown of the same, and their cost, at 10 per cent of the total project cost.[20] The contrast between the two also illustrates very well the differences between using local knowledge, local labour and local NGOs rather than importing technologies and equipment that cannot be sustained locally.

A feature of the government representation in UNEP is that the Governing Council representatives are usually different personnel from those who act as Permanent Delegates. In the CPR, erratic, or maverick behaviour is tolerated. Some individual PRs are strongly at odds with their governments, and so the arrival of different people, often from different ministries, to represent their government at the GC sometimes modifies or reverses positions taken by their countrymen in CPR. Only a small minority of PRs are scientifically trained, or even seconded from an environment ministry. Other countries send Foreign Office staff, or service UNEP as part of their bilateral staff accredited to Kenya. In the UK, the High Commissioner to Kenya is accredited as the Permanent Representative. In day-to-day terms, the officially-designated 'focal point' for UNEP is a First Secretary. He is primarily responsible for the UK bilateral aid programme in Kenya and in 1990 cited UNEP as one-third of his workload.[21] Some delegates to UNEP are also accredited to HABITAT, which shares the Gigiri office complex with UNEP. In 1990 there were 78 missions accredited to UNEP; in practice these are Embassy staff, working from the Embassy. Curiously, seven countries maintained embassies or consulates in Nairobi, but did not accredit any staff to UNEP. The Honorary Consuls of the Holy See, Lebanon, Liberia and Luxembourg may be reasonably excused for having their minds on either higher, or other things, but the decision of Iceland, Malaysia and Saudi Arabia appears unfortunate. The British, Japanese, German and Dutch, for example, have high-level representation, but also a high workload because these are missions with a busy, bilateral load of relations with Kenya; also Nairobi is often used for other East African representation (Uganda, Burundi, Rwanda, Sudan, etc.). In contrast, some active missions in UNEP frankly have very little else to do.

UNEP has been remarkably free of the overt political disputes that damaged the reputation of several UN agencies during the 1980s. A senior British respondent cited just four instances of conduct within UNEP that in his opinion could have been construed as politicisation. The issues cited concerned debates in the Governing Council:

● Ritual Soviet objections to the inclusion of West Berlin in FRG statistics, etc.

- Resolutions concerning Israeli water-management practices in occupied territories.
- Resolutions concerning the impact of apartheid on black agriculture in South Africa.
- A resolution on the invasion of Kuwait at the Governing Council, Special Session of August 1990.

Berlin concerned a representational matter, and a pretty elderly one at that; the other three might be considered extraneous, but given the necessarily broad scope of an environmental mandate, the differential impact of land-tenure policies on black agriculture in South Africa, and the role of both riparian rights and ground-water in the intricacies of the Middle East, these two appear central to UNEP concerns. The last was to be the basis of much more unambiguously environmental concern when in March 1991, Iraq commenced the systematic destruction and firing of Kuwait's oil-wells and the large-scale deliberate discharges of stored crude oil into the Persian Gulf.

Any suggestion of politicisation rests on the argument that the Security Council, rather than UNEP, is the place to pursue environmental crimes of this kind, or that the membership was guilty of selectivity and hence double standards in singling-out South African rather than, say, estate practices in Central America, where landlessness creates similar environmental impacts, as the peasantry are confined to marginal land. A clearer case of objectionable behaviour, inappropriate to the UNEP mandate, occurred in 1985 and centred on the question of nuclear weapons doctrine at the height of the second cold war. The 1985 Report of the Executive Director was subject to a vote when brought to the UN General Assembly for adoption. The crucial factor was Soviet pressure to include language on the risks of nuclear war and peace education. The report was eventually adopted by 149 votes to 0 with 6 WEOG abstentions.

Over a period of twenty years UNEP acquired its unique catalytic role within the UN system, and generally received accolades for the efficiency with which this role was discharged. However, the burden of coordination is different. UNEP is still a 'small secretariat' isolated from the principal organs of the UN system. It remains subordinate to the General Assembly. It is overawed by the budgetary resources of the larger specialised agencies and even its sister programme, UNDP. The organisational culture of the UN system is based upon autonomy and resistance to centralisation. The treatment of issues on a sectoral basis reflects this culture of autonomy, as does the separate existence of the specialised agencies for nearly fifty years, and the weak coordination of the several UN programmes. Sectoralism is anathema to the demands of sustainable development, which requires the cross-sectoral integration of numerous activities. UNEP could no more be expected to 'coordinate' the system-wide activities of the UN than could a medieval monarch 'coordinate' his feudal barons. The preparatory process that was initiated after 1989 to plan UNCED turned in its latest stages to consider the institutional reform of the UN system, so as to infuse the system with the principles of sustainable development. The reform of UNEP featured in these plans, and eventually endorsed the expansion of the catalytic role and the transfer of the coordinating role to a new organ created by UNCED, namely the Commission on Sustainable Development. Before discussing those reforms, it is necessary to consider more broadly the issues and limitations of environmental diplomacy which UNCED sought to address.

# 5 Two Cheers for Rio, 1992

This chapter has two major purposes. The first is to explain the origins of the United Nations Conference on Environment and Development convened at Rio de Janeiro, Brazil, in June 1992. This explanation will explore the inherent difficulties and limitations of that process, typical of any large multilateral conference, and thus illustrate the ideas of issue-linkage identified in Chapter 1, and continue the discussion of the UN role after Stockholm, described in the preceeding chapter. This discussion will also describe what might be called 'the path through the paper', namely the pressures exerted on the agenda of UNCED during its passage through the machinery of the UN system. This discussion is intended to demonstrate that any reasonable analysis of the prospects for UNCED would have emphasised prudence and played down the hyperbole which attached to the conference. This section concludes with an examination of the complex and conflicting nature of the many *national* attitudes to the UNCED agenda.

The second section of this chapter briefly summarises the achievements of the conference by reference to the several binding, and some non-binding Conventions, Declarations and Statements which were adopted. A more substantial discussion then addresses the far-reaching, and frankly surprising breakthrough proposed for the UN role in Chapter 38 of *Agenda 21*. This section includes discussion of the Commission on Sustainable Development, and of the proposals to strengthen UNEP and to release it from the more obvious contradictions of its dual mandate which were identified in the preceding chapter.


ORGANISING UNCED

(i) **General Remarks**

The Trail arbitration of 1939 clearly established the principle of international obligations arising from transboundary air-pollution incidents. Subsequent landmarks in environmental diplomacy included the International Convention on North West Fisheries, The Final Declaration of the Stockholm Conference, 1972, The Economic Commission for Europe Convention on Transboundary Pollution, 1978, The UN Convention on the Law of the Sea, 1982, and the Vienna Convention, Montreal Protocol and London Agreements on the protection of the ozone layer. Almost as a subplot, environmental advantages have followed incidentally from the implementation of certain arms-control treaties, most obviously the Partial Test Ban Treaty, 1963, the Outer Space Treaty, 1968, the ENMOD Convention, and the Non-Proliferation Treaty. This trend is discussed by Detter de Lupis.[1]

In this larger context, UNCED may be regarded as a follow-up to the Stockholm Conference convened 20 years earlier. UNCED not only sought to revive the cross-sectoral perspective of that conference, but also made explicit the political, organisational and scientific *linkages* between environmental degradation and economic development. The scope of the UNCED undertaking also represented a political reaction on the part of many governments to some very specific environmental *causes célèbres* which arose during the late 1980s. A series of accidental and negligent actions, such as those occurring at Bhopal, Chernobyl and Basel, radicalised and directed what had previously been a general unease. The rising pace of environmental awareness was evidenced by the rise of the Green parties in Europe, especially in West Germany, and by sophisticated NGO campaigns on issues such as nuclear testing, toxic dumping and whaling at the international level. Finally, the growing scientific consensus on stratospheric ozone depletion and the enhanced greenhouse effect, brought, after 1988, even the more quiescent Anglo-Saxon political communities to an unprecedented level of environmental political consciousness.

Substantial analyses of UNCED can be found in works by Grubb *et al.*,[2] and by Thomas.[3] The analysis offered here concentrates on those factors which acted upon the government delegations, empowered in the conventional diplomatic manner to agree conventions, declarations and statements of principle, more or less binding on their governments as each agreement allowed. The other aspects of UNCED, the 'tree-of life', the presence of Zen masters and the role of numerous entertainment personalities may strike the individual observer as either profoundly moving or risible, or somewhere in between.[4]

### (ii) A Linkage Too Far?

UNCED explicitly sought to connect the two agendas of environment and development. That said, the connection was not welcomed by all participants. To some observers the connection might have appeared spurious or opportunistic. However, a series of very genuine scientific connections exists between some issues of environmental degradation and the absolute poverty which characterises the lives of the world's poorest one billion people: for example, the use of wood for fuel by people too poor to enter the cash economy for kerosene, its consequent pressure upon marginal land and hence the advance of desertification. At the macroeconomic level, the inhibiting role of Third World debt in relation to the adoption of sustainable development patterns, rather than export-led growth, has been demonstrated.

Arguments concerning the *quality* of development in the North complicated the linkage. Dispute centred on whether Third World population growth or First World lifestyles were the greater cause of environmental degradation. The argument was to recur at all stages of the UNCED process. Concerning the global commons, the greater harm, both currently and historically, is clearly attributable to the latter. For example, the industrialised countries' rates of carbon dioxide emissions, CFC releases, and the rate of consumption of non-renewable oil and gas reserves, also of *non-ferrous* metals, both now, and especially reckoned cumulatively through 200 years of industrialisation fuelled by imperialism, substantially outstrip Third World consumption over the same time period, despite the latter's larger population. In terms of their *future* regulation, the linkage between these issues is therefore one of equity and burden-sharing between North and South. Equity between *generations* is implied in the adoption of *sustainable* development. It is not just a matter of the developed countries practising sustainable development for the sake of *their* children's children, but for the sake of this current generation of Third World adults and *their* children.

A further dimension of issue-linkage was strictly political. The Rio conference represented an excellent opportunity to harness the neglected 'cart' of the development agenda to the fashionable 'horse' of the environment agenda. After a decade of frustration and disappointment during the 1980s it was an opportunity the Third World would not let go. The Third World delegations sought action on a number of issues, most obviously debt relief, action on desertification, technology transfer, non-discriminatory environmental practices and standards, improved terms of trade, preferential access for semi-manufactures, and compensation funds for the best available technologies. These would be linked *politically* (however good or bad their scientific rationale) to the industrial countries' urgent agenda of climate-change, rational forestry-management, the preservation of biodiversity and provision for toxic waste management.

The Third World was divided within itself, most obviously between oil-exporters and oil-consumers on the question of deep cuts in carbon dioxide emissions. Too dependent upon oil revenues to stand cuts in global oil-consumption, and too rich to qualify for aid and assistance, the Saudis, and others, were to prove as obtuse as George Bush on signing the Framework

Convention on Climate Change (FCCC) as eventually negotiated. This issue will be developed later, in discussing national attitudes to UNCED. Other linkage strategies, such as that favoured by the Malaysians, sought to connect compliance with the FCCC and non-discriminatory application of controls on deforestation. Malaysia specifically threatened to withhold signature of the FCCC, if the tropical timber-exporting countries were stigmatised in the forestry agreement.

UNCED therefore carried the hopes and fears of a diverse and to some extent mutually hostile set of actors. UNCED therefore met all the theoretical conditions of 'complex interdependence' associated with Keohane and Nye and discussed in Chapter 1. UNCED was a process of negotiation in which three characteristics dominated:

- UNCED addressed a complex agenda (with no clear ranking or hierarchy of the more important issues, and much linkage between them).
- UNCED required numerous actors other than governments to be involved, (e.g. the UN, state or regional authorities, regional organisations such as the EC, transnational corporations and NGOs were all involved not only in the negotiation of agreements, but also in follow-up and compliance procedures, especially in the provisions of *Agenda 21*).
- UNCED recognised the nearly irrelevant role of military force as a credible bargaining threat or sanction for non-compliance with environmental agreements. (It is hard to imagine which problem of environmental quality could be improved by threat or use of force. Some future Security Council might wish to order the destruction of a rogue-CFC plant, operating in defiance of the Montreal Protocol but the thesis is inherently improbable on both political and scientific criteria.)

The conditions of complex interdependence described above (and elaborated in Chapter 1), created the characteristic features of negotiation associated with large multilateral conferences:

- A preference for adopting agreements by consensus (compared to majority voting, which may achieve a numerical but hollow victory, without the support of major actors).
- A preference for combining 'package deals' or 'issue-linkage' between items on the agenda (compared to the rationalist's preference for addressing each item on its own merits, unaffected by concessions on other issues).
- Asymmetrical or differential costs to the negotiating parties. (Some parties can wait longer than others. Parties frequently discount environmental quality at different rates, e.g. Brazilian compared to Norway on

forestry. Some, frankly don't care as much as some other parties depending upon the issue.) In short, the slow pace of the large multilateral conference favours those participants which are least in need of urgent attention to their particular problems. This is more likely to favour the North over the South.

● Consensus procedures reward the most intransigent. Whereas majority voting punishes those who hold out as an unappeased minority of one, the need for consensus rewards the intransigent party which can deny the completion of consensus. Peter Sands has referred to this as the 'slow-boat rule'.[5]

The convoy or conference can only make progress at the pace preferred by the slowest member. Furthermore, it is likely that the majority will go a disproportionate distance towards accommodating the intransigent party. Witness the attempts that were made to bring the USA and UK into the scheme of common-heritage provisions for deep-sea mining in the UNCLOS crisis of 1981–82. Here, however, the prisoner's dilemma strikes again. If it serves the rational self-interest of one party to hold out for exceptional demands, then it is in the interests of all to behave likewise. Issue-linkage will therefore tend to reward dogmatists, insomniacs, ideologues and the procedurally awkward. This is a confrontational style that offends the rationalist and will punish the pragmatist, the easily tired, the boy scout, the girl guide, and the martyr.

These negotiating characteristics are familiar in the European Council and, of course, are also reminiscent of the UNCLOS process in the years 1970–82. UNCLOS involved the negotiation of a Convention that ran to 320 articles, negotiated during 585 working days, over a span of 15 years. It would be unpopular, but prudent to suggest that the follow-up to UNCED, especially the implementation of *Agenda 21*, the completion of negotiations on forestry principles and elaboration of quantitative targets and a time-scale for the FCCC may extend over a similar period. Although easily criticised, negotiation at the UN level, in the style described above is unavoidable because conflicts of interests, special pleading, asymmetry of impact, and differential costs are central to the linkage between development and environment. As established in Chapter 1, a global international organisation is the only forum in which states can make agreements which are

● multilateral (beyond bilateral/regional solution),
● public (subject to scrutiny and verification), and
● simultaneous (to overcome the prisoner's dilemma and free-riding),
● legally binding (ideally, but not necessarily).

Conventions, treaties and protocols *are* legally binding, but a major debate in the preparatory phases of the Rio conference concerned the preference of some parties for non-binding agreements (as eventually applied to forestry), and for so-called 'soft-law', i.e. normative statements which may create a ratchet effect encouraging standard-raising behaviour. Sands, French and others also correctly emphasise the possibilities for regional conventions and progress, especially on issues such as regional seas, fragile, mountain eco-systems, and shared river-resources. Regional arrangements also have advantages in terms of balancing the needs of acceptability and credibility in verification and enforcement procedures.

Any system of verification has to strike a balance between a culture of cooperation that encourages participation, and a level of intrusiveness which is rigorous and gives confidence to other parties. This balance may be more readily attainable within smaller and culturally homogenous regional pacts. The principal argument against regional action, even if scientifically and environmentally sound, is that regional self-regulation may be distrusted by other groups and pacts, and especially by countries neighbouring any regional group. The parallel to arms-control verification is instructive, if comparing the global scope of the NPT to EURATOM, or to a hypothetical Middle-Eastern nuclear-weapons-free zone that would require Israel, Iran and Iraq to exchange and accept inspectors.

### (iii) Advantages of the UN Method

The adoption of UN mega-conference procedures has been shown to be both practically and conceptually unavoidable. Is there virtue in necessity?

- *Pacta sunt servanda*, the Latin doctrine that treaties are binding on those who sign them (and therefore only binding on those who do concur), carries a powerful warning against proceeding by way of the easily assembled, overwhelming majority. This practice *may* stigmatise a small minority of states left outside the consensus. For example, a draft convention on climate-change which won the support of all except the USA, the Russia Federation, India and China might have been adopted by 176 to 4 at UNCED, but it would have served little purpose. The USA and Russia alone constitute 45 per cent of current anthropogenic carbon dioxide emissions. India and China are the largest Third World contributors to greenhouse gas emissions. A treaty without them would address less than half of the problem.
- Complex agendas mean that give and take is possible, but only when all the issues and the demands linked to them have been made explicit (i.e.

no one makes unilateral concessions). This compels transparency, and limits the tendency to play a 'wild-card' or to work to a hidden agenda. This requires that the parties move from rhetorical positions, which in the marketplace of ideas are free and therefore plentiful, to fully-costed statements of priorities which, being expensive, are correspondingly scarce. This may create a third beneficial characteristic of the 'slow-boat' conference.

● Complex agendas mean that a holistic, or integrated rather than a sectoral treatment is possible. Not only is this desirable on environmental grounds, it is also central to the meta-linkage of environment and development. The Brundtland and Bertrand reports also suggest this direction as appropriate for UN structural reform.

UNCED was therefore convened as mega-conference in which despite the rhetoric of seeking one-world solutions to one-world problems, the so-called international community in fact confronted a massive, complex and potentially divisive agenda. To obtain the degree of consensus necessary it would, ironically, be necessary to

● heed the very asymmetrical interests of the parties,
● consider regional opt-outs, exceptions and procedures,
● recognise that even so-called framework conventions, drafted vaguely to gain consensus must contain specific obligations ('The devil is in the detail'),
● devise procedures for amendment of agreements on the basis of best available scientific knowledge.

This last point again demonstrates the parallel between environmental agreements and arms-control agreements. New technologies can overwhelm old treaties. Automatic provisions for review and amendment must be built into the original texts. Those well-versed in the ways of multilateral conferences might have predicted some of the shortcomings of UNCED. In particular, it was reasonable to suppose that of the 'jewels in the crown', i.e. the anticipated conventions on climate, biological diversity and forestry would encounter difficulties, as the complexity of the factors identified above would limit the chances of comprehensive agreement. As will be shown shortly, these predictions of partial and deferred progress were largely fulfilled.[6]

Having argued thus far that UNCED was conceived on too ambitious a scale, that it might have attempted to go a 'linkage-too-far', the contrary thesis, that in some crucial respects, UNCED did not attempt to go far *enough* is also tenable. Parallel to the UNCED process, throughout this period, were the negotiations of the Uruguay round of the GATT and other discussions on financing the IDA and structural reform of the World Bank and International Monetary Fund (IMF). Strictly speaking, UNCED without a trade agreement and net-new money (without unacceptable conditionality) was destined to be a hollow procedure. The multi-billion dollar trade-flows that might have been released by concluding the Uruguay round far exceeded the monies available 'on the counter' at Rio, in enhanced official development assistance or funds for harmonisation and compensation associated with the framework convention on climate-change or the Montreal protocol. Yet, there is a contradiction between some developmental and environmental aspects of the free-trade question. Free trade in agricultural products would in fact destroy many Third World producers with their low-energy input farming methods. Nitrogen-deficit farming in the North, combined with fantastic levels of price-protection in the USA, EC and Japan have combined to create a system of agriculture that is both environmentally inefficient while only profitable in the crazy world of 'set-aside schemes' and CAP subsidies. African countries, in particular, noted the double standards of European and American neo-conservative governments calling for the liberalisation of trade while using the tilted playing-field of massive agricultural subsidies to defend the agricultural vote in their own countries. A system of subsidies designed in the 1950s to ensure that Europe never suffered the ravages of hunger and blockade associated with the 1940s was, by the 1990s, being touted as a modest subsidy to preserve a rural way of life. If the British miner's union President, Arthur Scargill, had suggested paying miners a subsidy on their surplus coal to preserve 'a way of life' he would have been denounced as insane, or worse, a socialist.

### (iv) The Path Through the Paper

The UN's interest in environmental questions was consistent, if low-key during the 1980s. The General Assembly had, as a result of resolutions adopted in 1983 and 1987, produced an *Environmental Perspective to the Year 2000*.[7] This document was described by UNEP Director Mostapha Tolba as 'a broad framework to guide national action and international cooperation for environmentally sound development' (p. i). It combined a number of sectoral and global issues. As an indicator of rapidly changing priorities, it did not directly address climate-change, but discussed this issue only incidentally through its remarks on energy consumption. The UN's second System-Wide Medium-Term Environmental Plan (SWMTEP), for 1990–95 was developed in cooperation with the International Institute for Environment and Development (IIED). This document further demonstrated the acceleration of the climate-change question through the agenda of environmental diplomacy. The plan addressed a wide range of issues: atmosphere, water, terrestrial ecosystems, coastal regions, oceans, lithosphere, human settlements, health and welfare, industry and energy, peace and security and environmental assessment, management and awareness.[8] For each programme both general objectives and implementing agencies were identified.

General Assembly resolution 44/228 adopted in December 1989 called for a second conference on the scale of Stockholm, which would specifically adopt the logic of the Brundtland report and thus link the previously separate agendas of environment and development. The United Nations Conference on Environment and Development (UNCED) was scheduled for the first two weeks of June 1992, and Rio de Janeiro, Brazil, nominated as the host city. Very late in the preparations, however, the date was set back by four days to accommodate the Islamic festival of Eid. The UNCED agenda was developed over a period of two years by a Preparatory Committee (abbreviated habitually to 'Prepcom'). The provisional agenda was developed in collaboration with the Designated Officers on Environmental Matters (DOEM). This small secretariat, chaired by UNEP, brought together the programme officers in every specialised agency and UN programme with environmental responsibilities. In June 1990, Prepcom and DOEM produced what amounted to an agenda for research and called for technical papers, to be produced by UNEP over the next two years.[9]

As the draft agenda for Rio emerged, it read as an agenda clearly led by environmental issues to which a series of developmental objectives had been attached. This was reminiscent of the Stockholm process, which required consistent lobbying by the developing countries to maintain the developmental perspective.

I     The protection of the atmosphere by combating climate change, depletion of the ozone layer and transboundary air pollution.

II     Protection of the quality and supply of freshwater resources.

III     Protection of the oceans and all kinds of seas, including enclosed and semi-enclosed seas, and of coastal areas, and the protection, rational use and development of their living resources.

IV     Protection and management of land resources by *inter alia* combating deforestation, desertification and drought.

V     Conservation of biological diversity.

VI     Environmentally sound management of biotechnology.

VII  Environmentally sound management of wastes, particularly hazardous wastes, and of toxic chemicals, as well prevention of the illegal international traffic in toxic and dangerous products and wastes.

VIII Improvement of the living and working conditions of the poor in urban slums and rural areas, through eradicating poverty, *inter alia* by implementing integrated rural and urban development programmes as well as taking other appropriate measures at all levels necessary to stem the degradation of the environment.

IX   Protection of human health conditions and improvement of the quality of life.

X    Developmental driving forces, population, energy and industrialisation.

XI   Cross-cutting issues and integrative mechanisms for sustainable development.

*Source: UN General Assembly*, A/CONF. 151/PC/6. 27th June 1990.

Derived from UN General Assembly resolution 44/228, the provisional agenda for UNCED reflected the attempts to forge the linkage between environment and development discussed above. The provisional agenda passed from detailed specificity on some issues to a very general level of abstraction in its concluding items. It was itself the product of multilateral bargaining and compromise in the Prepcom process. Item VIII, in particular, suggested a Pythonesque committee-statement on the meaning of life. Repeated references to the importance of 'protection' were gradually supplanted by references to 'management'. The WCED (Brundtland Report) inspired concept of 'sustainable development' appeared at the last, suggesting an attempt to square the circle.

The Prepcom met for over two years prior to the convening of the plenary conference, in which all the major legal instruments were drafted. Little was left to surprise. However, as late as the last Prepcom in April 1992, key aspects of both the forestry and climate conventions were not subject to prior agreement, and so provided the central disputes to be resolved at Rio. Four sessions of the Prepcom were held: in August 1990 at Nairobi, March–April 1991 and August 1991 at Geneva, and finally at New York 2 March–3 April 1992. The sessions of Prepcom therefore totalled 13 weeks of activity spread over 21 months. Prepcom divided its work into three Working Groups, reflecting an attempt to divide the agenda on functional, or sectoral lines. On any day of the Prepcom it was common for two Working Groups to be operating simultaneously, thus considerably stretching the ability of smaller delegations to be in two places at once.

- Working Group 1 considered: atmosphere, land resources, biological diversity, biotechnology.
- Working Group 2 considered: oceans, freshwater, hazardous wastes.
- Working Group 3 considered: institutional and legal matters.

Working Group 3 only got into its stride late in the second week of the session at Geneva in August 1991, and contained all the debates concerning the reform of UN machinery. These questions and the financial dimensions of the agenda received further, late, attention at New York just three months prior to convening the plenary conference.

The visible UNCED comprised the Plenary Conference convened in Rio on 4–16 June. It was an intergovernmental conference for the most part conducted at the ministerial level, but attended in its final stages by an impressive number of Heads of Government. UNCED also sought to include a very large number and range of NGOs in its deliberations. In some cases (USA, Canada, UK), NGOs were admitted to the delegations. This may be viewed variously as the open door of pluralism or the genteel Anglo-Saxon preference for co-option at work. The NGO decision to convene a parallel or alternative conference, the Global Forum of 7892 organisations, 25 miles distant from the inter-governmental sessions, ensured their marginalisation.

### (v)  Climate on the Fast Track?

The climate-change question emerged in the late 1980s as an environmental issue of particular significance to western electorates and governments. The issue received a separate, fast-track treatment in a number of UN-related negotiations. Just as the Stockholm conference had been galvanised by the question of acid rain, so climate change came to dominate the environmental aspect of the UNCED preparations. While the advantage of focusing upon one dramatic issue clearly enhanced both popular and media comprehension of UNCED, the climate issue was itself a deeply divisive one, both within the group of industrialised countries, and between the industrialised and developing countries. Tough intervention on climate-change was feared by the South as a general distraction from the developmental part of the UNCED agenda. It was also claimed by some to be symptomatic of an eco-colonialist conspiracy to *restrict* Third World development on the pretext of preventing environmental degradation. These arguments are explored below.

As early as 1978, UNGA Res. 42/184 of 11 December had highlighted climate-change and biological diversity as two priority issues (paras 6, 7).

Two joint initiatives of UNEP and the World Meteorological Organisation (WMO) were created in the late 1980s, the Intergovernmental Panel on Climate Change (IPCC) and the World Climate Conference (WCC). IPCC presented its First Assessment Report at Sundsvall, Sweden in August 1990. The WCC was held during November 1990 at Geneva. IPCC operated in three separate working groups. It was open to participation by all UN members, predominantly expert in composition and although producing its Sundsvall report, was then mandated to continue, as an open-ended group.

The First Working Party (UK chair), examined the basic science of the enhanced greenhouse effect and came to a firm consensus on the likelihood of unprecedented global warming over the next century attributable to anthropogenic causes. The Second Working Party (USSR chair) discussed the environmental effects of climate-change, itself subject to extensive prior interest in the UNEP desertification programme. Neither working party produced conclusions that were in themselves politically controversial.

The Third Working Party (USA chair), on 'response strategies', considered central political questions raised by climate-change. The implications of measures necessary to reduce fossil carbon emissions to such a degree and within such a time frame as to be effective were disputed and costed. The interested parties which emerged in the Third Working Party revealed the extent to which positions were beginning to harden on the climate-change question. The significance of this division was the adoption of a Framework Convention on Climate Change (FCCC), identified by IPCC, the Ministerial Declaration of WCC and the UNCED Prepcom as the central objective of the 1992 conference.

Both the Special Session of the UNEP Governing Council (August 1990) and the Second Climate Conference Ministerial Declaration, mandated the creation of a group of legal experts to draft an FCCC. The so-called Ad Hoc Working Group was therefore brought into being. This group's first meeting at Geneva, on 24–26 September 1990 was attended by 71 states; a further 16 international organisations (both IGOs and INGOs) attended. The Geneva meeting adopted a timetable of four projected negotiating sessions, the first of these in early February 1991, at Washington, DC. In the period of sixteen months between the Washington meetings and the UNCED, governmental opinion in the USA and, to some extent, the UK, began to backtrack on the urgency of the climate-change issue. As the implications of intervention were increasingly recognised as *dirigiste*, and possibly anti-competitive, the US and UK balked at making firm commitments to reduce greenhouse-gas emissions. The UK was a latecomer to the '1990 by 2000' formula. The USA shunned it all the way to Rio with substantial implications for the content of the Framework Convention on Climate Change as it eventually emerged from the conference. This is dealt with below. So, although climate-change received a fast-track treatment, it was also negotiated quite outside the UNCED Prepcom process, so allowing the recalcitrant parties to claim that UNCED itself was not the time or place to conduct negotiations on the climate-change issue.

## NATIONAL ATTITUDES TO UNCED

### (i) The Activists

A small group of countries with highly developed environmental policies constituted the most active and innovative actors within the UNCED process. Some countries were predictable in their concerns; others much less so. Denmark, Germany and the Netherlands formed a cohesive group of European Community states prepared to make leading commitments especially on climate questions. They were joined by Austria, Canada and Norway among leading OECD activists. Malta, the Maldives, Mexico, Sweden and the UK were widely credited with specific initiatives (such as Malta and Canada on the Montreal Protocol, Sweden on financial commitments and UN reform, and the UK for its earlier climate-change advocacy, admittedly less obvious by June 1992).

A group of African states were determined to raise the status of desertification within the UNCED proceedings. Among the poorest of the poor, these Sahelian countries, confronted by absolute devastation, knew better than most the integrated nature of the environmental catastrophe they faced, as climate-change, land-use patterns, deforestation, grazing and population pressures, civil war and refugee-flows had each contributed to the dilemma they faced. They suffered most from the fragmentation and sectoralism of the agenda, despite rhetorical gestures towards the unifying concept of sustainable development. Discussion of desertification was postponed from the February 1991 Prepcom to the August meeting, and within that meeting postponed again for want of documentation. Prominent among these 22 Sudano-Sahelien states as activists were Chad, Ghana, Senegal and Mauritania. A typical example of the linkages between issues was seen in these countries' attempts to link discussion of forestry issues *with* action on desertification.

## (ii) The Ambiguous

In any UN context, states can operate in more than one role. This is in fact typical of large countries with long, complex and sometimes contradictory environmental agendas. Japan, because of its very understated financial role and energy-efficiency record, possesses great potential for influence, which is rarely exerted. While being a model of energy-efficiency, it is a rapacious consumer of imported tropical hardwood and has received international condemnation on issues such as hunting marine mammals and its reluctance to guarantee dolphin-friendly tuna-fishing techniques. (No tuna fishery is *tuna*-friendly.) Across the range of the UNCED agenda, Japan thus occupied an ambiguous position. For different reasons the USSR and Eastern Europe also occupied a curious position within UNCED. (The USSR was succeeded by the Russian Federation in the Prepcom after 1 January 1992.) As the full extent of the environmental degradation of these countries became apparent after 1989, and the wholesale economic restructuring on market-led lines was initiated, planning and emission controls on the process of reconstruction were potential casualties of the rush for growth. It would depend on the extent to which potential donors, such as the newly-formed EBRD, would apply environmental conditionality to their loans policies. The Africans and Latin Americans also regarded Eastern Europe as the cuckoo-in-the-nest among those countries seeking aid and technology transfer. For many economic and political reasons Eastern Europe is more likely than the Third World to attract inward investment and the transfer of clean-up technology on preferential rates. Eastern Europe is nearer, and its transboundary pollution problems directly impact upon Western Europe. Private sector investment is probably safer there than in Africa. Voyeurism and magnanimity towards recent enemies combine to favour the already well-favoured. The creation of the EBRD and the success during 1990–93 at attracting inward investment from Germany and Japan in particular support this view.

The UK moved so as to occupy similarly ambiguous ground. Britain has gained credibility for its positions on climate-change and the Montreal Protocol, signified by Mrs Thatcher's inclusion in Dr Tolba's personal list of the Global 500 'Roll of Honour and Environmental Achievement'. (Not many would be so brave as to place the former Prime Minister in any list with David Bellamy and James Lovelock, or indeed with the UNESCO Club of the University of Yaounde.) However, the UK occupied only a middle-ranking position within the EC12 on carbon dioxide targets, and appeared to have no legislative framework to achieve even the modest target of '1990 by 2000', while on a range of transboundary pollution issues –

ocean-dumping of raw sewage, power-station sulphur dioxide emissions, burning of chemical wastes, and radioactive reprocessing releases – the record of the last decade was as public as it was shaming. The late conversion to the '1990 by 2000' camp was typical of this headline-driven embrace of the issues.

## (iii) The Defensive

Saudi Arabia and, by extension, the other OPEC countries stood to lose oil-export revenues from the adoption of carbon dioxide emission controls. The Saudis and other Gulf states were also less likely to be concerned by desertification than many. However, their relative prosperity meant that they could not credibly make claims for financial compensation or for technology transfer at preferential rates. The Saudis were notably difficult at the final Sundsvall sessions of the IPCC in August 1990. In this the OPEC countries were of one mind with the USA. Brazil was determined not to be scapegoated on the question of tropical forestry policy, and led the campaign to widen the debate on forestry from *tropical* forestry to embrace *all forestry*, that is, to include the sustainable management of boreal and temperate forests as well.

They were joined in this by Malaysia, which denounced western attempts to apply blame to the Third World for both deforestation and population pressures. Malaysia also protested its innocence on the question of indigenous peoples' rights in relation to commercial logging, defending its treatment of the Penan people, and made formal proposals prior to Rio for a 30 per cent forest-cover minimum standard for each country. Malaysia offered to respect a 50 per cent forest-cover standard as a figure it could work *down* to if the northern industrial countries would plant *up* to the 30 per cent target.[10] Malaysia may have placed a radical tongue in its very commercial cheek, combining denunciation of western eco-imperialism with market-led, right-to-growth rhetoric. The political point, not lost on the Europeans, was that they had largely denuded their native temperate forestry, in the search for national economic development, and now sought to lecture the tropical timber countries, to 'do as I say, not as I have done'.

Brazil, India and Mexico spoke for their continents when they sought to force the linkage on this issue to a number of financial questions, such as debt rescheduling, debt-for-nature swaps and the Global Environmental Facility (GEF) developed by UNEP, IRBD and UNDP. GEF will be discussed in more detail in Chapter 6. Disputes centred on its size, its conditions for loans and its democratic accountability. Not unreasonably, the North wanted GEF to be accountable to its creditors. Equally unsurprisingly, the South wanted GEF to be accountable to its clients. India had the most rhetorical position on a number of environmental questions, especially those related to technology transfer. China has the capacity to break any carbon dioxide emissions agreement, as well as sharing certain Indian attitudes on the question of historic compensation.

The USA was largely in a defensive position at UNCED, especially on climate-change, bio-diversity and financial questions. The USA was keen to avoid being stigmatised on the question of primary energy use, and also keen to deflect the conference from *dirigiste* initiatives. The US favoured market-led mechanisms such as applying the 'polluter pays principle' through full economic costing, emission charges and tradeable permits rather than *quantitative* targets for carbon dioxide emission reductions. The US also sought to discuss all greenhouse gases, thus including reductions already agreed on CFCs, to delay the need for reductions in $CO_2$. On bio-diversity the US led a campaign to prevent retrospective claims by Third World countries to earn royalties from the commercial exploitation of indigenous plant species. The American argument ultimately turned on a variation of arcane legal objections to patenting discoveries, compared to inventions. US patent law pertaining to genetic material is already in a state of confusion. The executive branch of government was loath to take a firm view in international law on an issue subject to such dispute in domestic law.

These countries were therefore prominent critics of different leading objectives set by the UNCED process. While not formally 'veto-states', because there were no votes, they could withhold consent from the major agreements cited as the most important, and could attempt to link their assent to these agreements with concessions from other countries as part of a linkage strategy. For the South the priorities were clear: trade and debt reform, action on desertification, water quality, and reaffirmed rights to national sovereignty over natural resources, as well as demands for financial instruments which would have to bridge the two sets of objectives.

THE UNCED OUTCOME

The headline achievements of the UNCED plenary sessions included the adoption of a legally binding Framework Convention on Climate Change and a similarly binding Convention on Biological Diversity. The former was weakened by late concessions to American views. Despite the

European consensus on limiting carbon dioxide emissions in line with the '1990 by 2000' formula, the Americans insisted on deleting any quantitative or time-limited commitments. Instead, the convention uses the language of formulating strategies which aim to stabilise emissions. Again, the arms control analogy is instructive. Article 6 of the NPT obliged the nuclear-armed states to initiate 'negotiations in good faith towards a cessation of the arms-race'. It might be called the St Francis principle: 'Lord make me good, but not quite yet'. The FCCC also drew the hostility of several leading OPEC countries, including Saudi Arabia, Kuwait and Iran. This was reflected in the pattern of signatures, and the list of those countries, which six months later, had still not signed the FCCC.[11] The very partial nature of the FCCC meant that the negotiations continued after June 1992.

The Convention on Biological Diversity also attracted American hostility. It was not signed by the Bush delegation. The Administration objected to certain legal and financial implications, particularly compensation for commercial exploitation of natural products in the pharmaceutical industry. The United States revised its opinion in the following year after President Clinton's inauguration in 1993. In fairness, a number of other countries did not adhere to the Biological Diversity Convention, including some leading tropical forestry countries which cited their displeasure with western stigmatisation over forestry questions as sufficient provocation to withhold their consent from the Convention on Biological Diversity. The list of non-adherents included Cameroon, Singapore, Vietnam, Brunei, Equitorial Guinea and Sierra Leone, among prominent tropical forestry countries.[12]

The third 'jewel in the crown', which was anticipated during the Prepcom phase was the adoption of a legally binding agreement on forestry. In the end, no binding agreement was possible and a compromise document was adopted in the form of a non-legally binding 'statement of principles'. The full name of this document revealed something of the disputes the parties had faced. The 'Non-legally binding authoritative statement of principles for a global consensus on the management, conservation and sustainable development of all types of forests', hinted at two victories for the tropical timber exporters.[13] The document was non-binding (but 'authoritative'), and crucially did not limit itself to *tropical* issues, but also sought to embrace boreal and temperate forests, i.e. those of North America, Russia and Europe. One observer characterised the text as 'repetitive and regressive, clumsy and at times contradictory'.[14] The text affirmed the value of forests both in economic and environmental terms, and also identified the need for *international* action to meet the full costs of sustainable forest management. While affirming free trade principles, the text also made repeated references to the need for debt-reduction and financial assistance for countries pursuing unsustainable practices.

The conference also adopted the Rio Declaration,[15] a statement of principles similar in purpose to the Stockholm Declaration of 1972, and the massively detailed, path-breaking programme of action for the Rio follow-up, namely *Agenda 21* (i.e. an agenda for the 21st century). Less well-publicised was the agreement to initiate negotiations on a desertification convention and the adoption of separate but UNCED-inspired resolutions in the 1992 General Assembly on the protection of straddling and highly migratory fish-stocks, and the special needs of small island states.

The Rio Declaration, persistently referred to as the 'Earth Charter' in the Prepcom phase, was intended as a brief statement of consensus on the meaning of sustainable development. It was touchingly suggested that a text on one side of paper suitable for hanging on a child's bedroom was the original goal.[16] The Declaration in fact was longer than Maurice Strong's earliest wish. It also demonstrated, again, the manner in which two separate thought processes, that relating to environment, and that of developmentalism, met but did not marry. It is obviously tempting to compare the language of 27 principles adopted in the Rio Declaration with the 26 principles contained in the Stockholm Declaration of 1972. Despite some obvious similarities, the divergence between the two texts is substantial. The Rio text reflects the adoption of new language and concepts, most obviously that of sustainable development and the precautionary principle (Rio 3, 4, 5, 6, 7). The language used in these principles overcomes the naivety shown in the Stockholm text on the subject of renewable versus non-renewable resources. Sustainability is defined in the established manner derived from Brundtland, namely that developmental needs must be reconciled with environmental protection to meet the needs of both present and future generations. The special needs and priorities of developing countries are explicitly affirmed (Rio 6), and implicitly given priority in the concept of 'common but differentiated responsibilities' (Rio 7). Thompson notes,

> In plain English this means that although nations have a common responsibility to develop in an environmentally sustainable way, a greater responsibility lies with the developed nations. This was widely interpreted to mean that, because historically the developed nations produced most of the pollution causing climate-change and other global environmental problems, and because they are rich and have the most advanced technology, developed countries have a moral responsibility to help those developing countries who (sic) do not.[17]

Another explicit shift, but one of emphasis rather than language, is revealed in the attitude to national sovereignty. Whereas the Stockholm Declaration postponed the party-pooping affirmation of states-rights until paragraph 21, the Rio Declaration affirms national rights over resources in paragraph 2. Like its predecessor, the Rio text is qualified by references to the need to prevent transboundary pollution affecting either neighbouring states or areas beyond national jurisdiction; that is, the common-heritage areas of the high seas, Antarctica and outer space.

Subtle shifts include that from 'man' and his responsibilities, to those of 'states' (Stockholm 4, cf. Rio 7). Also, as might be expected, the economic language of export price support and economic planning favoured in the Stockholm text (Stockholm 10 and 14) is replaced in the Rio Declaration (Rio 12 and 11) by explicit references to the 'open economy', to management and to environmental impact statements. Population issues are well-buried in both, under the euphemism of 'demographic policies', and the rousing political denunciation contained in the opening paragraph of the Stockholm text has its much shorter and more literate analogue in the Rio text's affirmation of the rights of oppressed peoples (Stockholm 1, cf Rio 26). Warfare is denounced in both (Stockholm 26, Rio 24 and 25). Commitments to development are central to both, but significantly development itself advances from being characterised as 'essential' in 1972 to a 'right' in 1992. This language particularly offended the USA on the grounds that a right to development implied that other civil and political rights might be violated in the name of the right to development. Also new to the Rio Declaration is the explicit statement favouring the rights of women, youth and indigens, accorded very specific roles in *Agenda 21* and elsewhere (Rio 20–22).

*Agenda 21*, with over 2500 national and international policy commitments in over 150 programme areas, made by over 170 states, comprised the most specific undertaking of its kind ever negotiated. It provided for action at sub-national, national and international levels and, controversially, created two mechanisms for monitoring compliance which will be implemented in a so-called 'follow-up process' over succeeding years. One mechanism is the standard UN provision for a follow-up conference in 1997; the other is for national reporting procedures to be implemented and subject to examination in a new UN forum on an annual basis through to 1997. The latter has led to comparisons being drawn between the UNCED process and the work of the UN Commission for Human Rights. Both tasks, follow-up and national monitoring, fall to the newly-created Commission on Sustainable Development (CSD), which will be discussed in detail later.

*Agenda 21* itself is a remarkable document in many respects. Given that the formal plenary process at Rio placed so much emphasis upon the general and the abstract, the document is, for the most part, a very detailed set of recommendations. Its origins lie in successive UNCED Secretariat drafts and, although divided into 40 chapters, which might imply sectoralism *in extremis*, the document read as a whole is a blueprint of integrated policies for sustainable development. It addresses not only issues but constituencies, finance and institutional reform.

The comprehensiveness of the *Agenda 21* coverage can be inferred from the document's chapter headings (see Table 5.1). *Agenda 21* also contains a set of estimated costs for the implementation of the 40 chapter programmes over the seven years, 1993–2000. In the document itself, the figure cited is $600 billion *per annum for* the years 1993–2000.[18] The text did not flinch from recognising that

The implementation of the huge sustainable development programmes of Agenda 21 will require the provision to developing countries of substantial new and additional financial resources. Grant or concessional

Table 5.1   The contents of *Agenda 21*

|  |  |
|---|---|
| 2. International Cooperation | 23. Social groups |
| 3. Poverty | 24. Women |
| 4. Consumption Patterns | 25. Youth |
| 5. Demographics | 26. Indigenous peoples |
| 6. Human health | 27. non-governmental organisations |
| 7. Human settlements | 28. Local government role |
| 8. Policy-making | 29. Workers and unions |
| 9. Protecting the Atmosphere | 30. Business |
| 10. Land use | 31. Science and technology |
| 11. Deforestation | 32. Farmers |
| 12. Desertification | 33. Financial resources |
| 13. Mountain eco-systems | 34. Technology transfer |
| 14. Agriculture and rural development | 35. Scientific research |
| 15. Biological diversity | 36. Education |
| 16. Biotechnology | 37. Capacity-building in developing |
| 17. Oceans | countries |
| 18. Freshwater resources | 38. International institutional |
| 19. Toxic Chemicals | arrangements |
| 20. Hazardous wastes | 39. International legal instruments |
| 21. Solid waste and sewage | 40. The Data gap |
| 22. Radioactive waste |  |

financing should be provided according to sound and equitable criteria and indicators. The progressive implementation of Agenda 21 should be matched by the provision of such necessary financial resources. The initial phase will be accelerated by substantial early commitments of concessional funding.[19]

Of this truly global sum, £125 billion is supposed to be forthcoming in ODA from the industrialised countries and the balance of $475 billion per annum to come from public and private sources in the South.[20] The total figure represents approximately 66 per cent of annual global military expenditure, which was estimated at $900 billion by Brundtland and UNEP at 1985 prices.[21] The $125 billion which *Agenda 21* supposes will be forthcoming from the developed countries, 'would amount to no less than 1 per cent of the GNP of the developed countries'.[22] In view of these figures it is not surprising that Chapter 33 of *Agenda 21* concerning financial commitments was added very late in the negotiating process. Grubb *et al.* comment upon the $125 billion per annum,

> This figure is more than twice the current total disbursements of Official Development Assistance (ODA) from the developed to the developing countries. Coincidentally it is also close to the official UN target for ODA of 0.7% from rich countries. To this extent the financial estimates in Agenda 21 start to identify what this target contribution might be used for.[23]

The whole issue of current levels of ODA and the conversion of military expenditures will be more fully addressed in Chapter 6.


## THE COMMISSION ON SUSTAINABLE DEVELOPMENT

From the perspective of UN institutional reform, one of the most surprising and significant achievements of UNCED was the agreement to create a new high-level organ specifically to coordinate the implementation of sustainable development policies throughout the UN system. The remainder of this chapter will therefore focus on those provisions of Chapter 38 of *Agenda 21*, which contained explicit and far-ranging proposals for the creation of the new coordinating organ, and substantial proposals for strengthening UNEP. The Commission on Sustainable Development (CSD), will operate as a functional commission of the Economic and Social Council (ECOSOC). The Commission was formally established after debate in the General Assembly's 1992–93 sessions. It has been charged with the duty to reconcile the environmental and developmental programmes of the UN system. As will be shown later, this effectively lifts from UNEP an unreasonable obligation to perform a coordinating role beyond its modest means. The path to the creation of the CSD was complex. On the principle of 'physician, heal thyself', UNCED adopted, in Chapter 38 of *Agenda 21*, a series of measures for the overhaul of the UN's internal organisation. This was intended to achieve a shift in organisational culture from the sectoral approach which deals with environment and development as separate issues, to implementing sustainable development as a unified concept. The strong preference of the Anglo-Saxon delegations in the preparatory process was to achieve this objective without *either* creating any new organs, *or* by committing net additional resources. To the surprise of UN-watchers, inured to the 1980s Reagan–Thatcher atmosphere of 'no-new-agencies, no-new-money', the case for creating the CSD was made convincingly and successfully during the period after the March 1992 sessions of Prepcom in New York, and before opening the conference three months later.

The need for a new organ was mooted quite early in the UNCED process. Roddick cites the Brundtland Report as an early advocate, but Brundtland may have been thinking on more hierarchical lines.

There is a also a need for a high level centre of leadership for the UN system as a whole with the capacity to assess, advise, assist and report on progress made and needed for sustainable development. That leadership should be provided by the Secretary General of the United Nations Organisation.[24]

Later, Brundtland advised that the Secretary-General should create

under his chairmanship a special UN Board for Sustainable Development. The principle function of the Board would be to agree on combined tasks to be undertaken to deal effectively with the many critical issues of sustainable development that cut across agency and national boundaries.[25]

It is worth considering at length an earlier statement of purpose on creating the CSD. It demonstrates the full possibilities for structural reform envisaged in earlier suggestions.

One idea calls for the establishment of a 'Sustainable Development Commission' to which all United Nations bodies, agencies and programmes as well as 'treaty' secretariats involved in the area of environment and development would be accountable. It would meet annually and examine policies and programmes for promoting global action on environment and development and would be both a political deliberative body and coordinating mechanism for the United Nations system's activities in this area. It could also be convened on an emergency basis to deal with environment-related disasters and crises. Although such a Commission has been likened to the Commission on Human Rights, its exact terms of reference and its hierarchical place within the structure of the United Nations system remain to be defined.[26]

This maximalist statement was prepared by the Secretariat of the Prepcom in July 1991, nine months in advance of Rio. Although created to coordinate rather than duplicate the activities of UNEP and UNDP, the CSD may come to largely supplant the central purpose of ECOSOC itself. That may explain the late conversion of the Anglo-Saxon powers to creating the CSD. A question mark remains over the CSD–UNEP relationship. A substantial part of UNEP's success at maintaining donor-support from the G7 countries has been its modest catalytic and coordinating role. Commitments elsewhere in *Agenda 21* to expand the competence of UNEP raise questions about its size and tasks that will be discussed shortly.

The decision to create the CSD, and its role, can only be understood in the larger framework of the system-wide reforms of the economic and social organs of the UN initiated since 1988, and accelerated since January 1992. An element of fashion, as well as reform, can be detected in the creation of the CSD. Roddick suggests that the emergence of the CSD as a continuous follow-up mechanism for the UNCED, and one with extensive NGO connections can be traced to the Aspen meetings of July 1991. She cites the governments of Bangladesh, Norway and France as providing particular support. Other governments, especially the sovereignty-sensitive Malaysians, were opposed to the CSD's formation at this early stage.[27]

UNCED Secretary-General Maurice Strong proposed revamping the UN Trusteeship Council for the purposes of sustainable development, but a formal proposal for the creation of the CSD was made by Venezuela at the New York sessions of the Prepcom just three months prior to the conference proper. This received support from several African countries, Mexico, Netherlands, France and New Zealand, and the NGOs. Disputes continued concerning the chain of accountability, whether via ECOSOC, or direct to the General Assembly. India, China, Pakistan and Colombia led the campaign to restrict the CSD's monitoring and compliance procedures.[28]

Adam Rogers, otherwise rather sceptical of the UN role, described the CSD and Chapter 38 of *Agenda 21* in general as 'the most important part of this chapter, and a central pillar in the entire UNCED edifice'.[29] Besides the tasks of follow-up and national reporting, the CSD is also burdened with aspects of UN system-coordination. UN coordination shares certain characteristics with black holes. All who venture in are torn apart by forces beyond their comprehension, while no sight or sound of their struggles or screams ever emerges to deter others from following. It could be suggested that rather than being a dramatic move to implement sustainable development practices throughout the UN, the CSD is, in fact, being asked to go where, in turn the Administrative Committee on Coordination (ACC), the United Nations Environment Programme (UNEP), the Committee of International Development Institutions on Environment (CIDIE), the Designated Officers on Environmental Matters (DOEM) and even ECOSOC itself have been asked to go, more or less boldly, before.

The CSD's tasks are defined in Chapter 38 of *Agenda 21*. The Commission is a functional commission of ECOSOC, which was itself established by the Charter as 'the principal organ to coordinate the economic and social work of the UN'.[30] As a commission of ECOSOC, its membership is limited to 53. (There were suggestions at Rio to make the Commission a General Assembly organ and therefore potentially universal.) ECOSOC elects the 53 members for three-year terms, from a regional formula, giving Africa 13, Asia 11, Latin America and the Caribbean 10, Eastern Europe 6 and WEOG 13. The first elections were held on 16 February 1993, and shortly thereafter the first bureau was organised, namely the Chairman, and four Vice Chairmen, one from each of the UN regions. Razali Ismail of Malaysia was elected to the Chair, with Rodney Williams (Antigua and Barbuda), Hamadi Khouni (Tunisia), Bedrich Moldan (Czech Republic), and Arthur Campeau (Canada), as Vice Chairmen. In addition to the 53 member-states, any other UN member is able to attend as an observer. Furthermore, and consistent with the UNCED process, the CSD has adopted rules for NGO participation which are more generous than those usually operated by ECOSOC. In short, any NGO accredited to the UNCED will be able to apply for NGO status at the CSD, subject to review to be undertaken over the next two years on the whole question of NGO participation.

The work of the CSD will be organised over the five years running up to the UNCED review-conference in 1997. It has adopted a provisional agenda in which certain clusters of issues will be discussed annually, together with a list of issues to be worked through, several each year, thus covering the 2500 recommended national and international actions cited in the adoption of *Agenda 21*. Furthermore the CSD will receive the national reports of 189 UN members. Since the CSD is only expected to convene for a two-week period in advance of ECOSOC each year, it is clear that the bulk of its consideration will rest in the hands of its inter-sessional bureau and permanent staff assigned to CSD from the new Department of Policy Coordination and Sustainable Development.

The issues that will be addressed on an annual basis have been grouped as follows:[31]

A. Critical elements of sustainability; which will include international cooperation and domestic policy coordination, combating poverty, consumption patterns and demographic issues.
B. Financial resources and mechanisms.
C. Education, science and technology; including bio-technology, technology transfer, public awareness and scientific training.
D. Decision-making structures; including the integration of environmental and developmental planning (which might be described as the subtext of all sustainability questions, including the reorganisation of the UN's own capacities in these fields); also international institutional, and legal questions.
E. The roles of major groups; 'major groups' is UN-speak for the role of women, youth, indigenous peoples, NGOs, local authorities, workers and trade unions, business and industry, scientists and technologists and farmers. Each is addressed by a separate chapter of *Agenda 21*.

The clusters to be addressed on the rolling programme, probably starting in 1994 will be organised thus:

F. Health, human settlements and fresh water; a potentially massive issue including human health, the mega-cities and coping with their sewage. Despite the temptation to regard a variety of environmental issues in terms of the preservation of natural eco-systems, especially the more fragile and exotic among them such as tropical forestry, wetland, mountain and desert, the great majority of the *human* misery will be increasingly concentrated in the great cities. Twenty-two large cities are expected to reach a population of 25 million each during the 1990s, a list headed by São Paulo and Mexico City. Whereas New York, Paris and London each took over 150 years to grow to their 8 million size, Mexico City will add that number to its population in just 15 years.[32]
G. Land, desertification, forests and biodiversity; including integrated as opposed to sectoral approaches to the management of fragile eco-systems and the conservation of biological diversity.
H. Atmosphere, oceans and freshwater; the global issue of the atmosphere is joined with oceans, so defined as to include enclosed and semi-enclosed seas such as the Black Sea, and perhaps curiously in view of F

above, the protection of freshwater resources. The link is that riparian pollution, such as the Rhône, and Po is a major source of inshore marine pollution.

I. Toxic chemicals and hazardous wastes; this cluster includes the two politically sensitive issues of the illegal trade in toxic waste and the safe disposal of radioactive waste. In the case of the former buying a Third World country may be cheaper than buying land-fill sites, an obvious temptation to corrupt small sovereignties. The latter case concerns the choice between reprocessing and spent-fuel storage. Plutonium has a half-life of 24 000 years; no human system of government has maintained continuous law and order and record-keeping for more than perhaps 800 years on the most generous assessment of the Roman Republic and Empire.

Chapter 38 made a number of additional proposals for UN reform in pursuit of sustainable development. Reflecting the new ethos, the framework in which this is to occur stresses not only 'the principles of universality, democracy and transparency', but also 'cost-effectiveness and accountability'.[33] In addition to the creation of CSD, *Agenda 21* also provided for the creation of a High Level Advisory Board. The HLAB is envisaged as a group of experts, appointed in their individual capacity, not as representatives of their nationalities, and their selection does not depend upon any regional formula. Numbering between 15 and 25, the Board will act as personal advisers to the UN Secretary-General and will meet prior to the sessions of the CSD.[34]

## STRENGTHENING UNEP

*Agenda 21* also addressed the UNEP dilemma, which was identified in Chapter 4 of this study: namely, how to maximise UNEP's generally well-regarded reputation in the Herculean role of catalysis, while seeking to relieve the programme of the Sisyphean tasks of coordination. *Agenda 21* not only addressed the latter, through the provisions discussed here for the creation of the CSD and HLAB, but also addressed the former by seeking to enhance and strengthen the role of UNEP. However, eliminating the contradiction in this dual mandate by creating the CSD shifts the problem of coordination rather than solves it. UNEP will now become one more UN organ to *be* coordinated, *by* the CSD.

*Agenda 21* sets out a manifesto for growth in UNEP, although careful to use the financially reticent language of 'an enhanced and strengthened role'

(38.21). UNEP is specifically reconfirmed in its catalytic role, 'stimulating and promoting environmental activities and considerations through the United Nations system', also in promoting policies, and developing new techniques including activities in fields of accountancy and economics not currently strengths of the programme. UNEP is further charged with the dissemination of scientific research, and 'raising general awareness and action in the area of environmental protection with the general public, non-governmental entities and intergovernmental institutions'. (38.21 paras c–g). More specifically, Chapter 38 gives UNEP the lead responsibility for the development of international environmental law, environmental impact assessments, regional cooperation, technical advice to governments in collaboration with UNDP, and (a particular initiative of Dr Tolba's ) a role in environmental emergency planning (38.21 paras h–n).

Chapter 38 places other burdens directly on the UN Development Programme. UNDP, as the other half of sustainable development, in fact outranks UNEP within the system, by the criterion which counts, i.e. the member's voluntary donations. Taking the pre-Rio budgetary cycle, UNDP was in receipt of $948 million in 1988, compared to UNEP's meagre $31 million for 1987, just 3.2 per cent of the money donated to the larger programme.[35] UNDP's tasks are defined as having lead-agency status for capacity-building at local, national and regional levels. It is to mobilise donor resources, strengthen its own programmes and assist recipient countries in coordinating their follow-up to the UNCED (38.24 and 38.25). Other provisions in Chapter 38 assign further duties to the full range of appropriate UN agencies; UNCTAD, the Sahelian Office, the specialised agencies and the international financial organisations are each appropriately cited (38.27 to 38.41).

However, for sustainable development to become established as a culture-shift rather than an additional slogan, the UN system will require more than nominal commitments to *Agenda 21*. It will require a continuing overhaul of the mechanisms for coordination, and massive net additional funding. The lesson of the last decade is that whereas additional finance will definitely *not* be forthcoming without structural reform, it will *probably* not be available even when structural reform has been achieved.

In the period 1988–93, UNEP was subjected to extraordinary demands. The Montreal Protocol Secretariat, the UNCED Preparatory Committee calls for technical papers, the Joint UNEP/WMO work in IPCC, the Second World Climate Conference and the International Negotiating Committee for a Framework Convention on Climate Change (FCCC) each added to the workload. The 'small Secretariat' of about 250 professional staff is no larger than the science faculty of a medium-size British university. However, Chapter 38, and Tolba's own promotion of UNEP's growth during the last years of his period in office may create difficulties in addition to the more obvious opportunities. The generally very supportive attitude of the member-governments to UNEP has rested, until now, upon respect for its modest programme and good housekeeping, detailed in Chapter 2. The role UNEP has secured has been by virtue of resisting the inherently inflationary tendencies characteristic of UN agencies, compared to UNIDO, for example.

The continuing financial crisis within the UN could yet create pressures for wholesale mergers of functions and activities within the secretariat. Despite its favoured status in *Agenda 21*, UNEP, as a small programme, could still be vulnerable to asset-stripping. The small science programmes on desertification, coastal seas and climate change could be taken over by FAO, UNESCO and WMO, leaving UNEP with the Global Earth Monitoring System (GEMS) the Convention on International Trade in Endangered Species (CITES) and the International Register of Potentially Toxic Chemicals (IRPTC) as unique work well-performed by one agency in one place. (Although, in UN style, CITES is not in Nairobi or even Geneva, but Lausanne.) Given UNESCO's own problems of funding and image, it might covet the reputation of UNEP's work. The opposite scenario would involve UNEP being restructured as a specialised agency. UNEP, like UNIDO before it, might grow from a programme into an agency, with an autonomous budget derived from assessed contributions. The environment would then become yet another sectoral problem, something treated separately from health, education, food and labour, rather than the integrative solution to the problems of sustainable development.

Thirdly, and most in line with the *Agenda 21* recommendations, UNEP could be strengthened in its catalytic role, while retaining its essentially academic and informing role. Its informational, educational and catalytic roles can only succeed on the basis of proven excellence. To continue the university analogy, UNEP as a node of good science, funded to the tune of Caltech, MIT or Imperial could genuinely catalyse the international community. This is a task that UNEP can do. But 'coordinate' a system that was 25 years old before UNEP was created it cannot do, and should no longer be asked to do. The creation of the CSD recognises this.

CONCLUSION

It would be churlish to underestimate the role of UNCED in focusing public attention on the factors linking environmental degradation to the failure of development, and the dangers of overdevelopment. However, it would be unrealistic not to note the sense of disappointment which inevitably followed the levels of media hyperbole and governmental double standards which characterised those two weeks in Rio. On several criteria the conference's achievements fell substantially short of the goals claimed by the last-chance rhetoric which surrounded its deliberations. The conference existed in a complex *context* of multilateral diplomacy, with its origins several years ante-dating June 1992. That context also continues after the formal closure of the conference, both in UNCED follow-up procedures and the continuing pace of parallel negotiations on numerous issues, such as further tightening of the Montreal Protocol and the sucessful conclusion of the GATT round in December 1993.

Neither was UNCED the *only* contemporary negotiating forum for some of the issues it highlighted. The insistence of the USA on separating UNCED from the FCCC negotiations demonstrated a determination to maintain the diplomatic segregation of a typically cross-sectoral problem. This was not entirely self-serving. Although stigmatised at Rio for its aversion to quantitative targets for greenhouse-gas reductions, the US did have one leg to stand on, in seeking to insulate the continuing and complex FCCC process from the sound-bite diplomacy of the mega-conference.

Other claims for UNCED, especially those concerning its level of NGO participation, the role of indigenous peoples, of youth and what might be called the spiritual dimension of the conference, are enthusiastically described by Adam Rogers.[36] A critical perspective is offered by Peter Doran, who cites Mathias Finger's fourfold summary of the UNCED process.[37] He emphasises the extension of market-economic theories to environmental and developmental problems, the legitimation of the 'nation-state' in managing those problems, the effective marginalisation of NGOs and the co-option of 'the New Age model of politics, stressing individual rather than collective and local action'. Further, he alleges that UNCED elevated 'a planetary technocracy shaped by the demands of business, further eroding democracy and cultural diversity'.[38]

Two cheers for Rio. UNCED undoubtedly established sustainable development as a recognisable, if uncertainly defined concept in the lexicon of international relations. It established a totem to which at least verbal recognition must be extended in future negotiations. It also served to establish the linkage between environment and development, but fell far short of confronting the political and lifestyle implications of those linkages it did so well to reveal. UNCED might be said to have applied pluralist analyses and nostrums to a crisis which is now revealed to be more inequitable, more intractable and thus *structural* than UNCED's liberal architects and supporters might have wished. Many travellers on the road to Damascus have sincerely hoped that they would not be converted.

More important than the two legally binding but flawed conventions adopted at UNCED were the novel procedures for sub-national, national and intergovernmental review contained in *Agenda 21*. At the apex of these proposals are ambitious plans for an expanded UN role discussed here. As with all questions of UN reform there is no shortage of imaginative paperwork. The most important constraint is political will. The sole criterion by which to measure that will is by evidence of financial support from the G7 and OECD powers, which, despite the exigencies of the contemporary recession are, far and away, the most able to fund sustainable development. This can be achieved by embracing an expanded concept of environmental security that includes some diversion of conventional defence expenditures and an imaginative treatment of the debt burden. Only by these routes will sufficient multi-billion-dollar funding be forthcoming. No new taxes are popular, but the untaxed use of the commons has been shown by Hardin, Pearce and others to be responsible for their over-consumption and depletion – in some cases, beyond recovery.

Many things went unsaid at UNCED. That did not surprise the deep-green ecologist, and should not have surprised anyone with knowledge of the limitations of conference diplomacy. A comprehensive list of omissions, focusing on the challenge both to ecology and the equity of First World consumption patterns, would have provoked greater political disagreements. These are cited, with many criticisms, in Rogers' contribution.

The primary areas where UNCED needs more focus are; monitoring transnational corporations, enforcing the principles and conventions that were agreed upon, financing, setting the targets and timetables for meeting the objectives of the agreements, population, military [issues], the use of fossil fuels, debt issues, free trade issues, and deforestation.[39]

This chapter has discussed in particular the conference's proposals with the most immediate implications for UN institutions. UNCED did not directly confront two of the most pressing issues that have been discussed in Chapters 2 and 3 of this study, namely the impact of continued Third World indebtedness, and the lost opportunities for regulating and taxing the global commons. There is also a massive potential for UN structural reform outside the recommendations contained in Chapter 38 of *Agenda 21*. That synthesis will be attempted now.

# 6 Beyond UNCED: Revenues and Reforms

The secretariat of the Conference has estimated the average annual costs (1993–2000) of implementing in developing countries the activities in Agenda 21 to be over $600 billion, including about $125 billion on grant or concessional terms from the international community

*Agenda 21*, Chapter 33.18

External indebtedness has emerged as a main factor in the economic stalemate in the developing countries.

*Agenda 21*, Chapter 2.24

This chapter will first examine both the current limitations and future opportunities for the reform of ODA, the financial basis of UN operations, and alternative sources of revenues to finance sustainable development. It will secondly address the agenda of UN structural reform, that goes *beyond* the *Agenda 21*-derived possibilities raised at UNCED and described in the preceding chapter.

The United Nations is both the best and the worst place in which to conduct environmental diplomacy. It is the only arena in which all the world's states can meet on the basis of sovereign equality to negotiate new norms of behaviour and adopt binding conventions on a range of issues. It is also a place where representation is strictly limited to *states*. Entire nations, such as Kurds and Tibetans, and most indigenous peoples, such as Amerindians, aboriginals and pygmies, are excluded. NGOs, despite their self-importance, are marginalised. It is a place in which both political interests and bureaucratic structures favour the sectoral approach over the integrated approach to sustainable development. It is a place in which the gulf between declaratory standards and their implementation grows wider as many member-states' willingness to finance what they vote for, and to submit to verification and compliance procedures in conventions which they adopt, reveal the strains of confronting the full agenda of sustainable development.

As was seen in the preceding chapter, there was real progress after Rio in addressing the shortcomings and overlap of responsibilities within the UN system. Most obviously, the creation of the CSD represented a late conversion to the cause of creating an *additional* organ, after a decade of resistance from the US and UK leadership of the Geneva group of donors.

Since 1986 reform in the UN has been led by financial stringency, cost-cutting, zero real growth and an accountant's concern for cost-effectiveness. There is nothing intrinsically wrong with financial stringency in the organisation of international government. Responsibility for the dispersal of several billion dollars of official development assistance deserves care and cost-effectiveness. Problems begin when such stringency becomes a policy objective in its *own right* rather than a *measure* of efficiency in achieving a certain goal. By the late 1980s, the double standards involved in some attacks on the UN became substantial.[1]

The Reagan administration demanded and was granted consensus procedures within the Committee on Programme and Coordination. A further detailed review of the Secretariat's structure was undertaken by the Committee of 18 during 1986.[2] However, throughout this period, the US use of financial penalties against the UN became the single largest cause of the financial crisis faced by the organisation. After 1988 the source of the difficulties shifted from the White House to Congress, when successive bills introduced by President Bush to make good the arrears of several years foundered on Congressional opposition. The US used a combination of withholdings, late-payments and arrears, to become by the end of 1992 the largest debtor in the UN system. In September 1992, the extent of US pre-1992 system-wide obligations stood at $553.9 million, of which $266.4 million was owed to the Regular Budget. The cash-flow of the UN was also affected by the further $624 million due in 1992, that was still owing nine months into that year.[3] Despite the much-touted US claim to be unfairly carrying 25 per cent of the assessed budget of the organisation, the decline of the US contribution to multilateral programmes was so great as to rank only 18th in ODA per capita in 1992.[4] After 1990, the UN Secretariat was substantially reduced in staffing, and reorganised into larger departments, one with specific responsibility for UNCED, by a series of reforms initiated by Boutros Boutros-Ghali after February 1992 and completed by the end of that year. On 3 December 1992, the Secretary-General announced the creation of the Department for Policy Coordination and Sustainable Development under the direction of Under-Secretary General Desai. It is this Department which will in turn provide the staff for both the CSD and the High Level Advisory Group of individual experts. Although the CSD has been created to undertake the task of coordinating sustainable development policy, the CSD will in turn be subject to the coordinating efforts of ECOSOC and of the Administrative Committee on Coordination, which has in turn created a new sub-committee structure, the Inter-agency committee on sustainable development. Ultimately responsibility rests with the UN General Assembly, which is the principal organ to which all except the Security Council are accountable. Chapter 38 of *Agenda 21* confirmed the role of the Assembly as 'the highest level intergovernmental mechanism...the principle policy making and appraisal organ on matters relating to the follow-up to the conference'.[5] In addition to receiving the annual report of ECOSOC, the Assembly may also have to table an annual debate on the Rio follow-up, either in plenary, or in its Second Committee. The UN's way of streamlining itself can move grown men to tears.

The UN is clearly well-equipped for certain functions, and less well-equipped for others. An elementary distinction would highlight the following propositions.

● The UN system is well equipped for:
   (a) Collection, interpretation and dissemination of large-scale data sets, e.g. World Weather Watch, CITES, GEMS/GRID.
   (b) Consensus negotiations of norms; in the environmental field, examples of responsible environmental behaviour would include: common-heritage principles, demilitarisation of the seabed, the moon and outer space.
   (c) Consensus negotiation and adoption of rules in the form of treaty law, e.g. the Montego Bay Convention, the Vienna Convention, the Montreal Protocol, the London amendments, and IAEA conventions on nuclear-plant accidents.
   (d) Safeguards and inspection regimes to monitor compliance with treaties, e.g. the IAEA safeguards under NPT, and the compliance committee of the parties to the Montreal Protocol.
● The UN system is poorly equipped for:
   (a) Taxation and other financial powers creating incentives with regard to responsible environmental behaviour of its members.
   (b) Naming and publicising members in violation of norms and rules.
   (c) Imposing by vote, or administrative procedures, credible sanctions on violators.
   (d) Coordination of integrated programmes involving many agencies singularly engaged in sectoral work.

The demands of the environmental agenda have concentrated attention on the post-UNCED future for the UN's environmental responsibilities. The recurrent needs are for high-level coordination within the UN system, the integration of developmental and environmental planning and decisions, finding the appropriate role for competent non-governmental organisations and the problems of capacity-building in Third World states.[6] The underlying need is for the post-Cold War UN to consider economic and environmental security – that is, sustainable development – with the same degree of seriousness as military-security questions have received in the previous half-century.

The burden of Third World debt repayments eased slightly after 1992, as the effects of lower US interest rates, adopted during 1991, began to take effect. However, the impact of the lost decade, and of interest payments on rescheduled debts will persist. From an understanding of the problem of Third World debt, four discouraging propositions emerge. Firstly, so long as the level of indebtedness now carried, especially in Latin America and sub-Saharan Africa, persists, the environmental degradation of those regions will continue apace. Secondly, those countries most severely affected by debt will also degrade the common-heritage territories available to them. Thirdly, no obvious source of net additional capital is available to assist these countries; and fourthly, no general incentive exists for the outright forgiveness of debt by the creditors.

New capital can be found from new economic activities, and an incentive to relieve debt can be found that assists the protection of environmental quality, in terms that creditor nations can comprehend, in a spirit of enlightened self-interest, if official altruism is hard to come by in a recession. The ideas explored in this chapter focus upon raising international revenues by taxing the transboundary pollution of the commons. Most obvious among these is the introduction of carbon taxes that will act upon the release of carbon dioxide and hence delay and alleviate the trend towards to climate-change. The objective is to apply the polluter-pays principle, and so internalise one of the most obvious externalities of environmentally harmful economic activity.

Weizsacker and Jesinhaus observe that

> In a host of cases the polluter-pays principle cannot be applied in any strict sense because either repair is impossible (e.g. in the case of species extinction), or the damage is nearly impossible to quantify (e.g. damage as a result of the enhanced greenhouse effect), or it is impossible to apportion legal responsibility for pollution (e.g. when committed by companies which no longer exist).[7]

Other strategies to raise revenue from controls of greenhouse-gas emissions include the imaginative proposals for internationally-traded permits for levels of emission. These could be traded in a market and so reward those countries that reduced emissions, and sold surplus permits, and would impose costs on those that had to purchase additional permits up to their levels of emission. Other proposals such as those for taxing air-travel and levies on defence expenditure will be discussed as well.

The United Nations Convention on the Law of the Sea remains inoperative. Its provisions for generating revenues by licensing seabed mining activities in the area beyond national jurisdiction are therefore not yet activated. This postpones for the time being access to a substantial source of revenue for sustainable development. Less well-known provisions in the Convention require revenue-sharing between the coastal state and the UN in cases of oil exploitation beyond the 200-mile limit. So far only the Hibernia field off the coast of Canada could be included, but new developments off Ireland and Scotland might fall into this ambit. Other proposals to tax the commons, such as geostationary satellite orbit sales have been made. There is also the question of Antarctica. However, the moratorium on mineral exploitation agreed in 1991 has effectively ruled out this source. This failure to identify revenue potential in the management of the 'old' commons does not preclude their later development. However, what can first be done to enlarge the financial resources of the existing UN mechanisms? In particular, what flow of resources could be generated by the OECD countries if they actually implemented the 1970 pledge to donate 0.7 per cent of GDP for official development assistance?

THE UN FUNDING IMPASSE

At the close of 1992, two years after the declaration of a new world order which had placed the UN at the centre of efforts for conflict resolution, and six months after the Rio conference, the UN was in deep financial crisis. This crisis had been developing throughout the second half of the 1980s. In a report released in November 1991, Secretary General Pérez de Cuéllar highlighted the state of UN finances. As of 31 October 1991, unpaid assessed contributions to both the regular budget and the rapidly expanding peacekeeping operations totalled $988.1 million. Forty-nine per cent of this sum (i.e. $485.4 million), was owed by the USA alone.[8] By October 1992, the situation of US payments was little changed, and the UN's situation was worsening because of the non-payment of Russian contributions and those of other former Soviet republics. The Secretary-General's August 1992 statement revealed that other major debtors to the organisation's regular budget (that is, excluding peacekeeping arrears), included members listed in Table 6.1, each with an amount in excess of $5 million outstanding.

Table 6.1     Major outstanding UN regular budget contributions, 31 August 1992
(US $million)

| | |
|---|---|
| USA | 524 |
| Russian Federation | 138 |
| South Africa | 49 |
| Ukraine | 17.3 |
| Brazil | 33[9] |
| Yugoslavia | 8.2 |
| Iran | 7.5 |
| Turkey | 5.5 |
| Argentina | 5.5 |
| Total | 788 |

Other countries owing less than $5 million, but more than $1 million included Algeria, Belarus, Bulgaria, Chile, Cuba, Hungary, India, Israel, Libya, Mexico, Nigeria, Peru, Poland, Korea, Romania.[10] The United Kingdom's share of the assessed budget, 5.02 per cent or $49.4 million was paid in full by the time of the August review. Frustrated by reliance upon stopgap measures, such as borrowing from peacekeeping accounts to meet obligations under the regular budget, Pérez de Cuéllar had proposed in September 1991 a substantial series of measures to effect long-term reform. These were largely reiterated by his successor, Boutros Boutros-Ghali, in his Annual Report for 1992. Among the highlights of these two proposals were:

● charging interest on unpaid assessments, after 60 days,
● authorisation to undertake commercial lending,
● an increase in the working capital fund from $100 million, to $250 million,
● a new Peacekeeping Reserve Fund to allow immediate deployments rather than having to await separate votes on funding,
● a Humanitarian Revolving Fund raised by a one-off call for $50 million from the members.

The US reaction to these proposals, 'whilst not dismissive...was decidedly unenthusiastic about the prospects of implementation.'[11] The US objected in principle to levying interest charges on unpaid assessments, but maintained a more substantial policy preference for 'zero budgetary growth' within the UN. The position was reaffirmed in a paper circulated at the November 1992 opening of the General Assembly's discussion of the follow-up to the Rio conference. Arguing to maintain current budgetary levels, the State Department said,

> It is of signal importance that UN efforts in response to UNCED are firmly grounded on reliance on existing resources in the current UN budget. Activities beyond 1993 should be financed within a framework of zero real growth and maximum absorption of non-discretionary cost increases, taking into account savings achieved through on-going UN reforms.[12]

Twelve months later, On 6 October 1993, the Secretary General announced that the total sum owed to the organisation had in fact increased from the $988 million of October 1992 to $1700 million. This comprised $536 million on the regular budget, and a staggering $1200 million on peacekeeping accounts.[13]

Despite the rhetorical position of the Bush Administration on supporting the UN, this financial crisis in the organisation is in large part attributable to withheld payments, arrearage and late payments by the USA. Each of these terms is distinct and requires explanation. Withholdings are monies owed to the assessed budget of the UN, that is, the membership subscription, which the US specifically refuses to pay on the grounds that the US government disapproves of the use to which payments are put. Historically this has included monies adjudged to be assisting the PLO (e.g. educational programmes in UNRWA camps), and payments withheld from the UNFPA because of their use in China for abortion programmes. In October 1993 the US Congress began debate on a resolution to withhold a full 10 per cent of the US contribution to the regular budget of the UN until the organisation established an office of Inspector-General to oversee budgetary and staffing matters.[14] Arrears (or arrearage), are monies due from previous years, specifically more than 24 months late. Formally a member-state in this position can be denied its right to vote. 'Late payments' is more a judgemental than a legal concept. It refers to payments made unhelpfully late in the year they are due. Whereas some countries pay their UN contributions helpfully in advance, the US has, in recent years, taken to paying its annual liability partly in October, and then a final tranche, in the last week of December. A cash-flow problem results from the 10-month wait for the largest single contributor's money. In October 1993, the US made a contribution of $233 million towards the 1993 regular budget of the United Nations, and $358 million to peacekeeping accounts. These actions reduced the US arrears owed to the regular budget to $284.5 million, and peacekeeping arrears to $111.6 million.

Table 6.2     Status of US contributions to UN system, 1 September 1992 ($ million)

| | |
|---|---|
| *UN regular budget* | |
| Pre-1992 obligations | 266.4 |
| 1992 obligations | 298.6 |
| 1992 payments to date | 40.6 |
| outstanding to date | 524.4 |
| | |
| *UN peacekeeping* | |
| Pre-1992 obligations | 141 |
| 1992 obligations | 417.4 |
| 1992 payments to date | 349.5 |
| outstanding to date | 208.7 |
| | |
| *Specialised agencies* | |
| Pre-1992 obligations | 146.5 |
| 1992 obligations | 329.4 |
| 1992 payments to date | 31.6 |
| outstanding to date | 444.3 |
| | |
| *Grand totals* | |
| Pre-1992 obligations | 553.9 |
| 1992 obligations | 1 045.4 |
| 1992 payments to date | 421.7 |
| outstanding to date | 1 177.6 |

*Source: Washington Weekly Report*, XVIII-27, 11 September 1992.

Voluntary contributions to the UN system are handled under separate legislation from that covering assessed contributions for membership. Assessed contributions are part of the appropriation for the State Department, whereas voluntary contributions are funded under legislation for foreign assistance. Since 1992, some part of peacekeeping appropriations have been subsumed within Department of Defense expenditures. Table 6.2 shows the impact of late payments of the assessed contributions. The table shows the amounts of money due but outstanding on 1 September, with two-thirds of the Calendar Year complete. Michael Michalski, US advisor to the 5th Committee, argued that the impact of arrearage was exaggerated. Taking the middle 1960s as a base, he argued that compared to the $12 billion of cumulative assessments, 7 per cent remained unpaid.

Legislation to fully fund the UN regular budget for Calendar 1992 was not passed through both Houses of Congress until 1 October 1992. $225.8 million was paid to the UN on 9 October, with the remainder promised for early December. Even so, 'fully funded' meant withholdings of $16.4

million, and thus a contribution of $282.2 million. In addition Congress provided $92.7 million as the third of five instalments for the payment of arrears across the system. That said, the State Department still sought, in early discussion of the UN's 1994–95 draft budget, to maintain the policy of zero budgetary growth, associated with the most stringent days of the Reagan administration and the battles over budget and programme of the middle 1980s.[15] Twice during the year, in both August and September, Secretary-General Boutros Boutros-Ghali had had to resort to using money in peacekeeping accounts to meet payments chargeable to the regular budget. The situation in 1992 was exacerbated by the failure of the Russian Federation to pay its *circa* 9 per cent share of payments.[16]

Environmental questions, narrowly conceived in terms of the voluntary contributions to UNEP and other programmes, appeared to fare comparatively well in this scheme. The House of Representatives committee treatment of voluntary contributions highlights increases for UNEP, UNDP and provision for GEF, and reductions for the multilateral development bank (MDB) sector. The House vote on FY 1993 Foreign Aid Bill, introduced a $13.8 billion package, described by Congressman Obey, Chairman of the House Sub-Committee on Foreign Operations as the 'smallest foreign aid bill, as percentage of GNP, in the history of the country.'[17] House republicans justified reductions in MDB's budgets in environmental terms, being critical of the perceived big-project bias in their loan decisions. Foreign assistance requests from the President and level of appropriations approved by Congress, after 1 October broke down as in Table 6.3. From within the figure for International Organisations and Programmes, the environmental highlights included the amounts shown in Table 6.4. Attributing political significance to such small cash-sums may be uncertain, but clearly to the programmes concerned the difference between zero or $1 million for IUCN is significant!

Table 6.3    US funding for multilateral programmes

|  | 1993 request ($ million) | Congress passed ($ million) |
|---|---|---|
| Multilateral Economic Assistance | 14 072.407 | 13 897.275 |
| *of which* IMF | 12 313.857 | 12 313.857 |
| International Organisations and Programmes | 256.650 | 310.0 |

Table 6.4    US funding of UN environmental activities

|  | *FY1993 request* <br> *($ million)* | *Congress passed* <br> *($ million)* |
|---|---|---|
| UNDP | 124.0 | 125.0 |
| UNEP | 15.075 | 22.0 |
| ITO | 1.0 | 1.0 |
| CITES | 0.750 | 1.0 |
| WMO Climate Fund | 0.800 | 0.800 |
| IPCC | 0.300 | 0.300 |
| IUCN | 0 | 1.0 |
| Ramsar Wetlands Convention | 0 | 0.750 |
| Tropical Forestry Action Plan | 0 | 0 |

ERODING ODA

In round figures, the official development assistance given by the Development Assistance Committee members of the OECD is currently running at approximately half of the 0.7 per cent target. Despite donations by the small Scandinavian countries and the Netherlands, which well exceed the 0.7 per cent target, the total figure, and overall percentage of GDP donated is swamped by the effects of the much lower GDP *percentages* paid by the largest *gross* dollar-value donors, namely Germany, Italy, the UK and the US. The US is in the anomalous situation of never having agreed to the 0.7 per cent target in the first place, despite being frequently measured against it. Such are the trials of Empire.

If the 0.7 per cent target was implemented it would therefore effectively *double* the volume of ODA currently transferred from North to South, from *circa* $55 billion to *circa* $125 billion per annum. *Agenda 21* contained the crucial language:

> Developed countries reaffirm their commitments to reach the accepted United Nations target of 0.7% of GNP for ODA and, to the extent that they have not yet achieved that target, agree to augment their aid programmes in order to reach that target as soon as possible and to ensure prompt and effective implementation of Agenda 21.[18]

In the UK case 'as soon as possible' means when the Treasury permits, which may mean never. Less than six months after Prime Minister John Major's return from the podium at Rio, a 15 per cent cut in British ODA was widely discussed, having only just recovered from its historically lowest-ever level, namely, 0. 27 per cent of GNP, to a miserly 0. 34 per cent in 1991. When ranked in terms of how close to the 0.7 per cent target each country reaches, or indeed by how much a small minority of countries exceed this figure, some unusual names appear high in the rank order: see Table 6.5. The list is dominated by the Scandinavians, with one OPEC member, three members of the EC and only one member of the Group of Seven (G7) represented.

Table 6.5    Countries dispersing equal to or more than the 0.7 per cent target during the 1980s

| Country | ODA as % of GNP 1982–84 | 1987–89 | 1989 ODA, per cap. US$ |
|---------|------|------|------|
| Saudi Arabia | 2.7 | 2.4 | 80 |
| Norway | 1.0 | 1.1 | 217 |
| Netherlands | 1.0 | 1.0 | 141 |
| Denmark | 0.8 | 0.9 | 183 |
| Sweden | 0.9 | 0.9 | 212 |
| France | 0.7 | 0.7 | 133 |

The table of countries which do not meet the 0.7 per cent target is equally revealing. It contains six members of the G7. Some of the smaller and poorer countries of the Europe, such as Malta and the Irish Republic, contribute percentage shares of GNP which equal those of the UK and USA, and thus represent the widow's mite of the ODA.

Using the figures of the World Bank over the last five years of the 1980s, the size of the ODA shortfall can be determined. For the 18 countries of the development assistance committee, at current prices the shortfall of ODA compared to the 0.7 per cent target exceeded US $200 billion over the five years.

Table 6.6    Countries transferring less than 0.7 per cent of GNP target during 1980s

| Country | ODA as % of GNP 1982–84 | 1987–89 | 1989 ODA, per cap. US $ |
|---|---|---|---|
| Finland | 0.3 | 0.6 | 142 |
| Canada | 0.4 | 0.5 | 88 |
| Australia | 0.4 | 0.4 | 61 |
| Belgium | 0.6 | 0.4 | 70 |
| Italy | 0.2 | 0.4 | 63 |
| Libya | 0.2 | 0.4 | 27 |
| W. Germany | 0.5 | 0.4 | 80 |
| Japan | 0.3 | 0.3 | 73 |
| UK | 0.3 | 0.3 | 45 |
| Switzerland | 0.3 | 0.3 | 84 |
| Austria | 0.3 | 0.2 | 37 |
| Ireland | 0.2 | 0.2 | 14 |
| Malta | – | 0.2 | 15 |
| NZ | 0.3 | 0.2 | 26 |
| USA | 0.2 | 0.2 | 31 |
| Bahrain | – | 0.1 | 6 |
| Spain | 0.1 | 0.1 | 6 |
| Venezuela | 0.0 | 0.1 | 2 |
| UAE | 0.9 | – | 4 |

*Source:* Adapted from *World Resources 1992–93*, World Resources Institute (Oxford: Oxford University Press, 1992), pp. 236–7.

Table 6.7    The eighteen countries of OECD/DAC, GNP and ODA compared

| | GNP Current US $ billion | ODA Current US $ billion | ODA at 0.7% target | Shortfall US $ billion |
|---|---|---|---|---|
| 1989 | 13 950 | 46.7 | 97.65 | 50.95 |
| 1988 | 13 480 | 48.1 | 94.36 | 46.26 |
| 1987 | 12 050 | 41.6 | 84.35 | 42.75 |
| 1986 | 10 387 | 36.70 | 72.70 | 36.00 |
| 1985 | 8 490 | 29.4 | 59.43 | 30.03 |
| **Total** | 58 357 | 202.5 | 408.49 | **205.99** |

*Source: World Development Report* (Oxford: Oxford University Press, 1991), p. 240.

The performance of the United Kingdom is perhaps best summarised by extensive quotation from the Aid Report of *Christian Aid*, for 1991.

Christian Aid's annual Aid Report uses the government's own figures to analyze aid spending since 1975. The most recent figures available are for 1990. The following facts emerge as most significant: For the first time the overall flow of money was *from* Developing Countries to Britain. Repayments on past loans exceeded the total flow of money from the UK to Developing Countries – not just new loans but also export credits and investment. Debt repayments exceeded official and private flows by £2493 million.

Government aid spending as a percentage of Gross National Product fell to its lowest ever level of 0.27%. The UK is now fourteenth in the league table of donors. The Government has agreed in principle to reach the UN target of 0.7% but has yet to set a timetable for its achievement.

The amount of aid provided also fell last year, whichever measure is used. Public Expenditure on Aid fell by four per cent to £1725 million, a fall of 11 per cent after adjusting for inflation. Official Development Assistance fell by six per cent, 13 per cent in real terms. The value of Official Development Assistance in 1990 was only two-thirds [of] its 1979 value.

The value of aid to south Asia has been halved in the last decade. In 1990 Bangladesh received 50 pence per person from the UK Government. Cambodians were given only one pence each.

The downward trend in Government spending was not reflected in money donated to charities. In comparison the generosity of the public allowed aid spending from voluntary organisations to increase from £150 million to £184 million – an increase of 13 per cent after inflation. The amount of voluntary grants is now just over one-tenth of Government Aid.[19]

After the devaluation of sterling in September 1992, following the UK's exit from the Exchange Rate Mechanism, a public expenditure review was initiated which suggested a further erosion of the ODA commitment. In October, the Treasury proposed a 15 per cent cut in the programme. Oxfam director David Bryer was especially critical of the impact on long-term bilateral aid, which had previously benefited from the attention of successive Conservative ministers.[20] The Archbishop of Canterbury, George Carey, referred to the 'privatisation of morality'.

I do not underestimate the financial difficulties and political pain which the government faces. But if we think the going is rough here, let us remember the calamities unfolding in other parts of the world.[21]

## THE GLOBAL ENVIRONMENTAL FACILITY

The Global Environmental Facility (GEF), was launched in 1990, for an initial three-year trial as a result of cooperation between three agencies: UNEP, UNDP and the World Bank. Initially funded to the tune of $1.3 billion, GEF was to operate by a combination of grant-aid and low-interest loans. It was the focus of both favourable publicity and controversy during UNCED. The innovation and size of resources devoted to GEF was promoted by the G7 as evidence of the UNCED parties' commitment to sustainable development. In terms of net new monies there was not much else to point to at Rio. Despite early suggestions of a breakthrough in the level of Japanese contributions to GEF, very few solid pledges were received. Germany and the UK promised to 'play their part' in refunding GEF to a total of $2 billion.[22] Japan pledged a total ODA outlay of $70 billion over five years, with perhaps 10 per cent of this (just $1.4 billion per annum), being earmarked for environmental projects.[23] However, to the developing countries the size of the fund, and the number of caveats attached to its operation was evidence of the weak and partial commitment to sustainable development of which the G7 were accused. GEF gained an additional significance because other UNCED-linked funding initiatives were not implemented. The World Bank President called for funds towards an 'Earth Increment' of $5 billion in new pledges to the International Development Association (IDA). By December 1992, six months after the Rio conference, these had not been met. The tenth replenishment of the IDA was completed in that month. It raised 'about $16 billion over the next three year period, [and] represents little, if any, real increase in ODA funds'.[24]

GEF is significant, despite these limitations, because it is an attempt at integrated rather than sectoral action between three major UN organs, each traditionally jealous of their autonomy. Furthermore, it is specifically targeted at actions in defence of the global commons. This has also provided grounds for criticism, because funds are not available for local projects. Monies are available from the GEF for projects in four areas of environmental concern:

- limiting and reducing greenhouse-gas emissions,
- preservation of biological diversity,

- the protection of international waters, and
- the protection of the ozone layer.

Among the projects considered suitable are pilot projects for renewable energy sources, the reduction of gas-flaring in oil-producing countries, control of methane-release in coalmining and projects for protection of forests as carbon-sinks. Forestry protection is evident in projects for preservation of bio-diversity, schemes to counter desertification and to protect vulnerable wetlands. The international waters projects has targeted investment in waste-disposal facilities for polluted ballast that would otherwise be dumped at sea, and measures to reduce pollution in international rivers. GEF projects in relation to preservation of the ozone layer are restricted to countries which are party to the Montreal Protocol.[25] In fact the so-called 'ozone fund', also administered by the UNDP, UNEP and World Bank *troika*, was established *prior* to the creation of GEF, with commitments of $240 million.

GEF funds are only available to countries with per capita GDP of less than $4000 in 1989. The funds *are* additional to existing ODA, and must be applied to schemes which would not be economically viable for that country without support from the facility. In other words, it is in the nature of action to protect the commons that it is *non*-viable for any one country (because there is no property interest), and so it suffers neglect or abuse for want of collective action. GEF funds are also supposed to attract support from elsewhere:

> Favoured projects are those which most increase the leverage of GEF funds by mobilising substantial portions of total project costs from the World Bank, regional development banks and other financiers.[26]

Although preventive strategies for climate-change, ozone-layer protection and maritime-pollution issues are therefore eligible for GEF funding, *adaptive* strategies for climate-change, and the mass of issues central to Third World urban life, must continue to rely upon national or bilateral funding from traditional sources.

## CARBON TAXES

Carbon taxes may sound novel, but of course fossil fuels already attract high levels of taxation in Europe and Japan, although much less so in the USA. Carbon taxes seek to reduce the consumption of fossil carbon, i.e. of coal, oil and natural gas, and so reduce levels of carbon dioxide emissions, consistent with the polluter-pays principle. Carbon taxes would not only reduce consumption of fossil carbon through the incentive to greater energy efficiency, but they would also encourage a shift within the *mix* of fossil fuels, from coal to oil and from oil to natural gas. Although consistent with reducing carbon dioxide emissions and with overall energy-efficiency, this reverses the logic of husbanding *non*-renewable resources over the longest, *sustainable* time-period. Reserves of natural gas are more ephemeral than those of crude oil, which itself is surpassed many times over by the energy content of proven reserves of coal. A commonly assumed order of magnitude in the UK cites North Sea oil at 30 years of recoverable reserves, and British coal at 300 years. Recoverable reserves of 25 million tonnes of coal were sealed at Thornton, Fife, in 1992.[27] The UK government's announcement of pit-closures by British Coal in November 1992, emphasised the advantages, in terms of fuel efficiency, of the shift to natural-gas-burning power generation (the 'dash for gas'). They further, correctly, emphasised lower carbon dioxide and sulphur dioxide emissions. The longevity of the domestic coal reserves was a rearguard claim of the miners' union, invoking strategic and long-term options for energy policy. It did not find favour with the now-privatised electricity utilities, which would rather import cheaper coal from Colombia and South Africa than pay a premium for home production.

Beckerman suggests a number of reasons for the political appeal of pricing for pollution. The 1980s produced a political climate in the OECD that was generally anti-regulatory and yet also increasingly sensitive to environmental concerns. He suggests that this partly arose because the problems have become worse or at least more publicised, and partly because concern for environmental quality is more likely to rise in a more prosperous society. Reconciling the two trends suggests the use of market mechanisms to achieve goals which twenty years ago would have met with a *dirigiste* reflex. Taxes are generally anathema to advocates of the hidden hand, but Beckerman argues that taxes (and tax relief), are more acceptable than regulations, and also more enforceable. Taxation to correct the failures of the hidden hand, such as social costs, the externalities, etc., come IEA-approved.[28]

> If we could provide ourselves with *unlimited* amounts of clean air or water we would not mind how much of it was 'used up' by the polluter. But we become worried about it when the environment is a *scarce* resource so that, for instance, the pollution of a river or a beach means that we are deprived of enjoying its amenities. The key point is that the environment is a scarce resource and pollution is, in effect, a use of this resource.[29]

In Europe, especially in the UK, Switzerland and Italy, petrol is already heavily taxed. However, between European countries the pattern is uneven, as UK–Dutch disputes concerning the subsidies for glass-house consumption of natural gas, and the problems of VAT harmonisation in preparation for the 1993 single market have shown. The Europeans are already at the forefront of taxing motor-vehicle fuel, and already vary the rates of taxation for *other* environmental reasons, such as the lower rates levied on diesel and on unleaded fuel. Given the levels of fuel efficiency in car-design and levels of taxation on petrol already levied, it is perhaps surprising that the EC has been among the first rather than the last with carbon tax proposals but, as noted, other environmental advantages are cited, for example to assist in limiting acid rain. In general, carbon taxes would encourage investment in more fuel-efficient technology and import substitution, because, with the exception of UK oil and gas, and Dutch gas supplies, the EC members are wholly dependent upon imported oil and gas. Oil is predominantly a Middle Eastern dependency, natural gas-flows come from the Russian federation, limited LPG supplies are shipped from Algeria, and in the future new pipelines may be developed from now-independent Turkmenistan.

Japan faces very similar incentives to fuel efficiency. Japan is the world's second largest economy, with a GDP per capita comparable to that of the USA. However, this feat is achieved at a level of carbon emission per unit of GDP, almost *half* that of the USA.[30] This enviable situation lay behind the reluctance of the Japanese to approach the negotiation of a global convention on climate-change on the basis of percentage reductions in emissions. Such cuts would be comparatively easy for the profligate, and correspondingly difficult for countries such as Japan with historically low levels of emissions. On the contrary, fuel-efficient Japan would gain a double advantage from an internationally administered and set carbon tax, because it would be the United States that would bear the highest costs among OECD trading competitors, not only in meeting the taxes levied, but also in the impact upon comparative economic advantage. Not surprisingly, among 1992 presidential candidates, only Mr Ross Perot had the political courage to advocate a federal gasoline tax. He lost.

At the most modest levels such as £0.40/tC, that is 40 pence per tonne, or about 1 per cent of primary fuel prices, in the UK alone a figure of £300 million per annum would be realised and, worldwide, approximately £2 billion. In a much more ambitious proposal Grubb suggests an international application.

If a tax was to be applied at a level sufficient to alter industrial choices directly...it would have to be of the order of £20 or $30 per tonne of carbon. The revenues accruing to the fund from such a tax would be over $100 billion: some one hundred times the current registered budget of the UN. Even if the basic problems of control, administration and allocation of such funds could be overcome, it seems unrealistic to expect such contributions to be achievable.[31]

In spring 1992, as part of its preparations for UNCED, the EC Commission undertook an extensive programme of investigation for the introduction of carbon taxes. Described in the UK House of Lords European Communities Committee as a unilateral response to a global problem, the Commission's draft directive proposed a $3 per barrel tax on crude oil. This was proposed to rise during the decade to reach $10 per barrel by 2000. In effect, with oil prices at just under $20 per barrel at that time, the Commission was proposing a levy of 15 per cent on crude oil production. Although at first sight this might seem severe, it should be recalled that pump-prices of petrol had in fact *halved in real terms* over the period from the second oil-shock of 1979 to the recession of 1992. (Pump prices were at approximately £2 per gallon in both 1979 and 1992, having fallen substantially in the 1986 trough of OPEC pricing. In the same period the retail price index doubled.)

Carbon taxes do not only raise questions of international competitiveness. Proposing unilateral flagellation as an attempt to embarrass the Bush Administration before Rio certainly appeared overly optimistic. But then, this was just a small instance of how the political dynamic of volunteering for environmental protection becomes constrained in a period of recession, compared to the comparative virtue of proposing sacrifice in a period of growth. Also problematic during a recession (that is, a period of rising unemployment and pressure on welfare provision), are the intuitively regressive implications of any carbon tax. Heating, lighting and cooking costs form a greater proportion of household expenditure, the poorer the household.[32]

Carbon taxes may be designed which are revenue-neutral; that is, they are introduced as *substitutes* for other taxes on either income or expenditure and do not raise the overall tax-take. It is harder to conceive how carbon taxes might be devised which have *distributionally* neutral effects, if they are intended to make people do less of something that poor people, proportionately speaking, do more of. As if to demonstrate that this is a real and not hypothetical issue, the UK prepared during 1993 to attach VAT at 8 per cent on domestic electricity and gas supplies to take effect in April 1994, rising to 17.5 per cent in April 1995. The government claimed environmental grounds for this action. Only in the context of some elaborate voucher-system linked to welfare payments, could this regressive effect be avoided, in practice exempting the poor from the workings of the tax. The possibility that poor people might prefer to put on an extra woolly jumper, sell their electricity vouchers, and then buy more food, is the radical logic at the heart of the *internationally* traded carbon-emission permits idea, of which more shortly.

Michael Grubb has noted that the whole question of comparing generating efficiencies ignores a more pertinent option:

> The greatest flaw is to suppose that measures to limit greenhouse gas emissions need be costly. Many could in fact be economically beneficial. Energy efficiency is the most obvious example. Investing in generating plant at £1000/kilowatt while not investing in conservation at £300/kilowatt may be a consequence of current 'free markets', but it can hardly be described as good for the national economy.[33]

Japan, in particular, would seem to have benefited from the discipline of having no indigenous oil. J. K. Galbraith has argued that the absence of indigenous raw materials can be a blessing.[34] The heights (or depths), of imaginative market solutions were recently illustrated by the case of the 'cash for clunkers' swap agreed in California. The agreement with the state authorities allows US oil-refining companies to buy up old cars and scrap them, thus eliminating a calculated total of 6400 tonnes per annum of CO, HC and NOX emissions, and to offset this amount against delayed improvements in oil-refinery plant. A 20-year-old car in a poor state of maintenance may cause 90 *times* the pollution of a new model with electronic ignition and a three-way catalytic converter.[35]

## INTERNATIONALLY TRADED PERMITS

On the principle that turkeys never vote for an early Christmas, unilateral or even regional action to create a self-imposed burden of carbon taxes is unlikely. As Grubb has predicted, and the weakening of the Framework Convention on Climate Change at Rio demonstrated, there are also fundamental obstacles to international agreement on percentage reductions in carbon dioxide emissions. Depending upon whether emissions are measured in per capita terms, or by per unit of GNP, at present exchange rates, or per unit of GNP, adjusted for purchasing power, then respectively, the USA, China or the six former communist states of Eastern Europe can be stigmatised as the most profligate nations.[36] If, rather than measuring current emissions, a cumulative view of the last 200 years is adopted, in order to apportion responsibility for anthropogenic carbon dioxide levels, then clearly the northern industrial powers, including the former Soviet republics, appear even more culpable. On the other hand (and there is always another hand), as Paterson and Grubb's subsequent work has shown, when projected into the future, if China and India raised their per capita emissions to even the current European levels (*circa* 3 t/C per annum), this would, on the basis of their combined population of two billion persons, exceed current US emission totals by a factor of four times. This would reverse the impact of even quite severe emission-reductions in North America, Europe and Japan. The point is not academic. As the debate surrounding the Nordhaus evidence has shown, the United States government under President Bush balked at the costs of *preventive* approaches to climate-change, specifically severe emission controls compared to the costs of *adaptation* to the phenomenon. If it were true that those preventive measures would not only be more expensive than doing little or nothing, but would also be *negated* by the next half-century's economic and population growth in India and China, then the political pressure to do nothing increases. Neither the statistics nor their implications were changed by the election of the Clinton–Gore Administration.[37]

The problems associated with defining rates of emissions, burden-sharing and special pleading, led Paterson and Grubb, among others, to advocate tradeable permits for emissions instead of carbon taxes or negotiated national percentage reductions in emissions. In themselves, tradeable permits are not new. The vogue for 'vouchers' as a market-led option in education and the provision of public utilities, has been extant for fifteen years. In the USA the Environmental Protection Agency (EPA) pioneered their use for electricity utilities during the 1980s. The concept is simple. Their execution, especially internationally, is more problematic.

In essence, a tradeable permit system requires an authority, such as the UN or EC to set a global or regional figure for carbon dioxide emissions (or any other pollutant for that matter), and to issue permits for a certain level of emission, to the participants, on an agreed formula, such as per head of population. Thereafter, those actors (utilities, or countries), wishing to exceed their permitted quota of emissions must enter into the market and buy surplus permits from utilities or countries which by virtue of their poverty, or energy-efficiency or some combination of the two, have permits which are surplus to their needs. Heavy energy-users therefore acquire an incentive to economise internally rather than pay for permits. Meanwhile, the poor and the efficient have their reward either in cash, or maybe in the form of development projects, debt relief or some other environmentally desirable service. While this is an elegant proposal, the snag is obvious. There is no international analogue for national government with the authority to issue, and more pertinently, to verify and enforce compliance with the regime. Furthermore, the political arguments that could be predicted on the question of national targets, would be re-enacted over the question of the appropriate distribution of the permits. A straight per-head-of-population system would have obvious advantages to the poor. It also has redistributive justice to support it. The poor are least responsible on historic criteria for the last two centuries of carbon dioxide emissions, and currently, per head, inflict the least damage. Douthwaite argues passionately, if sentimentally, for the per capita solution.[38] Grubb offers an adjustment which would take some cognisance of the age-structure of each country, only permitting adults to count in the calculation, so as not to reward high rates of population increase.[39]

The advantage of tradeable permits over carbon taxes is that a system of internationally traded permits provides *incentives* to behave in the environmentally responsible direction without the burden of setting *targets* to be achieved. Grubb maintains, 'The primary aim of a system like this would be to decouple the international decision on acceptable levels of carbon emission from the tortuous process of trying to allocate those restrictions among nations.'[40]

Three objections to the idea of tradeable permits concern equity, verification and free-riding. To anticipate the criticism that permits would tend to accumulate in the hands of the richer countries, forcing the poorer to buy back permits to allow the development of their own economies, Grubb suggests that permits be reissued periodically. In effect, each issue would be valid for a certain period of time, perhaps five or ten years, and then recalled and a new issue made, restoring the parties to the level playing-field. The Biblical precedent discussed earlier would suggest a suitably sabbatical seven years. Greene argues that verification is a prerequisite rather than an optional frill, for *any* system of emission control. The precedent of the Treaty on the Non-Proliferation of Nuclear Weapons (NPT) is clear, and generally encouraging. The system of inspections and control, known as the 'safeguards system', and administered by the International Atomic Energy Agency (IAEA), has a generally excellent record of balancing credibility with acceptability in its operations. Until the discovery of the extent of Iraq's breaches of the NPT, no unauthorised diversion of nuclear materials had been recorded under the NPT regime since its inception in 1970. If states have shown themselves willing to accept such a degree of intrusion into such a highly-sensitive sector of their national economy as the nuclear fuel cycle, the lightness of hand necessary to verify and inspect carbon dioxide emissions from power stations, large industrial complexes and the financial and other records of power-generating companies or utilities, cannot be insurmountable.[41] Finally, states free-riding outside a tradeable permits regime could be isolated by imposing carbon duties on their imports. That would, of course, raise difficulties in the context of GATT. *Agenda 21* warns against unilateral actions being taken against countries on environmental grounds. 'Should trade policy measures be found necessary for the enforcement of environmental policies certain principles and rules should apply.' These are cited as non-discrimination, the least restrictive of trade, transparency, adequate notification and special consideration for developing countries.[42]

## TAXING DEFENCE SPENDING AND AIR TRAFFIC

Among Boutros Boutros-Ghali's 1992 proposals were two ideas which might be represented as attempts to raise revenues from intrinsically international activities: namely, levies on the international arms trade, and a proposal for a levy on international air travel. These would have the effect of raising resources for the UN that would be under the direct control of the organisation.[43] However, there a number of reasons to be sceptical of any proposals to tax the arms-trade. A levy on the arms-trade is a dubious proposal in terms of efficiency. The basis of the trade is largely secretive, its true volume is distorted by bribery and by counter-trade (that is, barter) to overcome exchange rate problems. Linking ODA to a fixed proportion of defence expenditure is more feasible, as are suggestions to convert some part of defence expenditures to activities which address the issues of environmental security. In terms of arguments for the conversion of military resources to development, the case is morally compelling. Military expenditure as a percentage of GNP ran at 5.4 per cent during the 1980s. One UNEP estimate of world defence expenditures, suggesting a figure of $1.6 million per *minute*, would fund the initial $1.3 billion of GEF in 13 hours.[44] Mahbub Ul Haq, of UNDP, has suggested that a annual reduction in defence expenditures globally of 3 per cent would yield $1500 billion ($1.5 trillion), in ten years. $1200 billion would come from the developed countries and $300 billion from developing countries.[45] These figures may be optimistic. An analysis of defence expenditures over the period 1985–91 shows that in real terms, i.e. at 1985 prices and exchange rates, the reduction in global defence expenditures since the peak of the Cold War has already been impressive. However, this reduction has not been linked or converted to increased ODA commitments.

Table 6.8    Defence expenditures, at 1985 prices ($ million)

|              | 1985    | 1991    |
|--------------|---------|---------|
| NATO         | 357 966 | 317 327 |
| USSR         | 241 500 | 91 621  |
| Other Europe | 25 187  | 19 768  |
| Middle East  | 68 559  | 67 592  |
| Africa       | 7 119   | 5 234*  |
| Asia/Oceania | 58 363  | 70 053  |
| L. America   | 11 567  | 10 289  |
| *TOTALS*     | *770 261* | *581 884* |

*Numerous African figures not available.
*Source: The Military Balance 1993* (London: International Institute of Strategic Studies), pp. 218–21.

In real terms, NATO area annual defence expenditure has therefore already fallen by $50 billion, over the six years which marked the decline and termination of the Cold War. The larger part of the global fall is attributable to the near $150 billion reduction in Soviet expenditures before the dissolution of the USSR. Defence expenditure in Asia actually rose, in real terms, during this period. The largest rises were recorded in China, India, North and South Korea, Japan, Pakistan and Taiwan.

There are some commonsense reasons for identifying international air-traffic as a potential source of international revenues for sustainable development, namely progressivity, fuel efficiency and equity.

● Measured on the global scale of 6 billion persons, only a very small number of the world's richest citizens ever travel by air. A modest levy on their ticket would represent a very gently progressive form of taxation, relative to the world's majority to whom a bicycle represents exotic transport.

● In terms of fuel-efficiency per passenger-kilometre, air transport is the least efficient means of transport compared to surface transport by car, bus or train (between two and four times more efficient). As a supplementary form of carbon taxing, an air-transport levy makes sense.

● Due to the very regulated nature of international air traffic (high levels of security and identity checks are already conducted to deter terrorism), it would be an efficient tax to collect by a ticket levy. This already occurs with earmarked levies for security procedures, in the USA, and the familiar airport tax levied apparently, for sheer devilment, by numerous countries. This is important, because ease of evasion is a major objection to any tax both on grounds of efficiency and equity.

## CARBON BANKING

Much of the focus of public attention in the west on forestry questions has centred on tropical deforestation. The point was forcefully made at UNCED that the extensive deforestation of the European and American temperate forests was largely accomplished three centuries ago. Meanwhile, the felling of the boreal forests continues in the Far North. This does not exclude those countries from responsibilities in the sustainable development of global forestry. During the UNCED process, a substantial effort to include the *potential* forest-lands of Europe and North America was mounted as part of the attempt to prevent the scapegoating of the tropical timber exporting countries. The boreal and temperate forests can in fact be substantially enlarged with several potentially beneficial results, some linked to concepts of carbon taxing.

As part of the uncertainty surrounding the question of climate-change, the notion of a comprehensive survey of sources *and* sinks of carbon dioxide was mooted within IPCC and UNEP. To understand the carbon cycle and the rise in atmospheric carbon dioxide, it is necessary not only to compute anthropogenic sources of carbon emissions, but also to compute the gains and losses among the sinks or stores of fossil carbon. The *political* connection becomes apparent when high-emission countries might offset their impact by enlarging their carbon sinks, principally by extending forestry. This focuses greater blame upon the rapidly-deforesting tropical countries, while reflecting virtue upon OECD countries which can practice sustainable forestry more easily. In the spirit of internationally traded rights to sources and sinks, high-emission countries might prefer to finance conservation or even reforestation projects abroad, and claim the 'sink' so preserved or created against their level of emissions at home. One source suggests that reforestation of tropical rainforests would have to extend to 3.9 million hectares per annum to absorb just 1 per cent of OECD emissions.[46]

The reform of the European Community's common agricultural policy (CAP), has been placed in motion by the McSharry proposals, a trend confirmed by the US–European GATT agreement of November 1992. A proposal for reducing the output of surplus commodities such as oil-seeds, which may be more constructive than set-aside, would be to adjust subsidies and incentives towards extending lowland forestry. 'Carbon banking', in the form of diversified broad-leaf and conifer plantation, would maintain rural employment and population, would serve to fix carbon dioxide efficiently and could be linked to a national system of carbon taxing, by offering carbon-tax credits for 'banking' carbon in the form of vastly extended forestry plantations. The central belt of Scotland has been mooted for one such scheme, combining the virtues of carbon banking with more down-to-earth gains in land-reclamation and employment-creation in a region with deep and obvious scars of 200 years of now-exhausted coalmining, oil-shale extraction and steelmaking.[47]

European forestry passed through several centuries of decline and clearance, due to the industrial demands of charcoal, iron-smelting, shipbuilding, pit-props and railway construction, as well as increasing agricultural land use. However, European forest-cover has in fact been *increasing* over the last half-century. Excluding the territory of the former USSR, forest-cover increased by 2 per cent between 1930 and 1960, and by 19 per cent in the period 1960–80.[48] A similar recovery of forest-lands occurred in the USA after 1930, partly in response to land-management practices and the creation of extensive public forest-lands; also, as in Europe, improvements in agricultural efficiency allowed the reversion of marginal land to forest-cover.[49]

This progress was, however, threatened after 1980 by increasing evidence of acid-rain damage to certain regions of European forestry and the north-eastern United States. The phenomenon which first attracted public attention in Germany as *Waldsturben* or 'tree death', varied in its impact from less than 1 per cent of trees with leaf damage in Portugal to 22 per cent of trees with leaf damage in Czechoslovakia. In all, over 18 countries were affected. The transboundary nature of the phenomenon required transboundary action, most obviously seen in the ECE Convention on Long Range Transboundary Pollution, and the debates between Scandinavian countries and the UK concerning culpability.

## DEBT-FOR-NATURE SWAPS

The name of this device accurately describes its purpose and method. Debt-for-nature swaps seek to create liquidity, ease the burden of debt and conserve the natural environment, all through one device. A debt-for-nature swap is a particular kind of debt-swap in which an outstanding foreign currency debt is redeemed on the secondary debt market for less than its face value, and the sum converted to local currency, which is then applied to some previously-agreed conservation scheme. The original debt may be purchased by a northern NGO, or, by a government as part of its ODA disbursements. It has been suggested that northern banks may wish to surrender or donate debts in this way, either because it would make them feel nice, or more probably because it would influence public opinion – aware and critical of the banks' practices and profits from Third World debt – and thus, at the very least, debt-for-nature swaps might constitute a cheap form of advertising for them.

Despite the appearance of creating a no-lose situation for all parties, debt-for-nature swaps have in fact only had a very limited impact upon easing the burdens of the most enthusiastic countries seeking to apply them. Described by Paul Vallely as 'small beer in the overall picture of Third World debt and not therefore worth further digression', such swaps were initiated in Bolivia in 1987, and have remained largely restricted to Latin American applications, in particular Costa Rica.[50] That first case illustrates the working of the idea very clearly. Conservation International, an American NGO, purchased $650 000 of Bolivian debt for the secondary market discounted price of $100 000. They then wrote off the debt in return for the Bolivian government's pledge to create a reserve of 1.6 million hectares in the Beni forest. In another case, Vallely describes the Midland Bank's write-off of $800 000 debt owed by Sudan. The Sudanese government then paid the equivalent sum *in local currency* to UNICEF for re-afforestation projects in Kordufan province. An investigation by the US General Accounting Office into the operation of debt-for-nature swaps revealed the very limited scope of application to December 1991. The total amount of debt swapped by 13 countries then stood at $126.4 million, or just 0.047 per cent of their total commercial debt. Costa Rica accounted for 68.4 per cent of the total swapped, with swaps to the value of $86.4 million. Ecuador's swaps, at $18.5 million, accounted for a further 14.6 per cent of the total. In the case of Costa Rica, the value of the debt-swap amounted to only 1.9 per cent of that country's total commercial debt.[51]

The costs and benefits of the approach are mixed for all parties. The NGOs may be involved in extensive bargaining to set up such deals. The GAO cited 18 months as normal. Debt-for-nature swaps may therefore prove to be a long shortcut. The NGO must always balance these costs against the gains of a cash donation. As expected, the banks themselves could see little point.

Banking officials told us that the economic advantages to be gained from donations of debt holdings were insignificant, and that if they donated debt instruments for one country, pressures could build for expanding such treatment to other countries. Furthermore they said that the banks' shareholders generally did not support donating bank assets for charitable causes.[52]

Another, more general, problem concerns the inflationary implications of creating additional purchasing power in local currency, in countries that are frequently prone to inflation anyway. 'A debtor government may print money to finance the debt conversion and thereby increase its money supply.' The GAO note that several Latin American countries, including Brazil, suspended debt-swaps for just this reason.[53]

For these reasons the future role of debt-for-nature swaps appears very limited. The outright cancellation option, as practised by the first case cited, in Bolivia, avoids the potentially inflationary conversion to local currency. It does, however, rely almost entirely upon trust between the voluntary agency and the sovereign state each to keep its side of the deal. The alternative, by which property rights would be invested in the sponsoring NGO, is likely to offend the sovereignty of the host nation. Verification may be eased by indirect government involvement. As the financier of such swaps, the donor government can create the possibility of a formal treaty obligation on the host to maintain its conservation bargain. In the US some debt-for-nature swaps have been funded with money provided to the NGOs by the US Agency for International Development (USAID).

This concludes the review of fund-raising opportunities linked to the exploitation of the commons, and attaching to their exploitation a level of taxation, encouraging both the conservation of the commons and providing revenues for sustainable development. The final questions raised in this study concern the recognition of environmental security as part of sustainable development, and the further structural reforms of the UN, not envisaged in implementation of *Agenda 21*, that this might require.


CONCLUSION

Proposals to build up economic–development structures within the UN, alongside and of equal importance to its traditional Security Council structure, are not new. The possibility of co-joining environment, security and UN reform is, however, more apparent now than at any time previously. This is not just a reflection of the rising environmental agenda, but also a consequence of the potentially veto-free UN Security Council of the post-Cold War system. Using the concept of environmental security, it would hypothetically be possible to elevate the environmental agenda to the Security Council itself. A wide range of environmental quality issues, and their likely effects, may create substantial military security problems for states in the next century. The most salient of these is that climate-change, desertification and even very modest sea-level effects in territories such as Egypt and Bangladesh may create mass-refugee migrations and land competition. In the Middle East, groundwater-abstraction and river-water diversion and rate of abstraction will exacerbate territorial disputes.[54] Policing enlarged exclusive economic zones (EEZs), damages actions for transboundary pollution and radiological accidents may all feature in the future causes of deteriorating relations between states. Secondly, no amount of unilateral action in a realist framework will protect stratospheric ozone, or maintain constant sea-levels. Indeed, continued, damaging unilateral acts of pollution, dumping and $CO_2$ emissions will contribute to the creation of worse conditions, in a classic illustration of the prisoner's dilemma.

The role of the UN system in achieving levels of environmental security which meet these new possible threats is, as always, the function of any intergovernmental organisation to act as a forum in which governments can overcome the temptations of the prisoner's dilemma and negotiate public, simultaneous and binding commitments to improved standards of conduct and adherence to peaceful means of conflict resolution. Recent events in relation to deliberate oil-spillage in the Gulf and the destruction of oil-wells in Kuwait have shown that crimes against nature and ecocide may in the near future be promoted by some states (the Nordic countries and small island states, etc.) as the moral equivalents of crimes against humanity and genocide. Deliberate acts of commission or omission in respect of environmental security may come under the scope of Security Council competence as 'a situation which might lead to international friction, or give rise to a dispute' (Article 34). These might include transboundary pollution, radioactive contamination, and destruction of pelagic fish stocks, etc. UNEP's role in this would, in conjunction with the International Law Commission, be to develop the appropriate norms and customary international principles into codes of treaty-law. A start has been made in this field. In 1977 the Geneva Protocol I Additional to the Geneva Conventions of 1949, relating to the victims of armed conflict, created obligations for states to refrain from damaging the natural environment.

Care shall be taken in warfare to protect the natural environment against wide-spread, long-term and severe damage. This protection includes a prohibition of the use of methods or means of warfare which are intended or may be expected to cause such damage to the natural environment and thereby to prejudice the health or survival of the population. (Article 55)

Other expanded roles for the UN and its agencies might include extending the IAEA safeguards model into verification and compliance with environmental agreements. UN reform could be discussed within the radical framework proposed by Maurice Bertrand who, with touching faith in the EC, has proposed the remodelling of the UN Secretariat/UNSC on the lines of the EC Commission/Council of Ministers, thus granting more continental-style executive initiative to the secretariat. This is a non-starter, given the dominant US critique of the secretariat. The question of a European seat to replace those of France and the UK has also been mooted, and received support from President Clinton's Ambassador to the UN.

Given the French–UK veto over their own removal from the permanent five, a more credible scheme of reform is one that recognises and institutionalises the need for a G7–Third World consensus. An 'Economic Security Council' of just 15, comprising the G7 plus the world's eight largest populations, would represent approximately 71 per cent of world population and 75 per cent of UN-assessed contributions.

Table 6.9    A proposal for an Economic Security Council

| | |
|---|---|
| Asia | China, India, Japan, Indonesia, Bangladesh |
| Latin America | Brazil, Mexico, |
| Africa | Nigeria, |
| WEOG | USA, France, Germany, UK, Italy, Canada |
| E. Europe | Russian Federation, or CIS seat. |

Expanded to 24, for regional balancing, adding, say nine from Africa, Asia and Latin America on an elected rotating basis (as at present), and voting by two-thirds of assessed contributions and or by a two-thirds population requirement, would compel the need for a much broader consensus.

Can the post-Cold-War UN confront environmental security with the same degree of vigour and new thinking that has accompanied the shift towards new and imaginative actions in the field of peacekeeping? Not only has the scope of such operations been extended, but new skills, such as humanitarian relief, election supervision and the quasi-imperial reconstruction of civil society, as in Somalia and Cambodia, have been added. Environmental security – that is, actions to protect the eco-system from threats to its sustainability – deserve intellectually and financially a similar degree of attention.[55] The financial crisis of the UN peacekeeping budgets shows that the costs of making war can be more readily summoned than the costs of maintaining the peace. Any attempt to extend the concept of security into preventive actions concerning the preservation of the environment will be even harder to establish. Whereas $9 billion were raised for the 1991 war in the Gulf, largely by subscription from the non-combatants Japan and Germany, the UN was owed $800 million of its $3 billion peace-keeping accounts at the close of 1992. The Secretary-General has suggested in *Agenda for Peace* that some part of the near $1000 billion in current prices expended on defence globally might be converted to the purchase of environmental security. This requires a paradigm shift in how societies define their security. The Secretary-General cited mass-migration of refugees, population pressures and barriers to trade among the causes of 'poverty, disease, famine, and oppression', which create insecurity and tensions between states.[56]

As has been shown, the range of very credible threats to ecological sustainability is growing: desertification, international water-rights disputes, food security, the prospect of mass-migrations of refugees caused by environmental degradation and the disruption of economies and lifestyles are not fanciful, but fit with the low-medium scenario for climate-change over 50 years predicted by the IPCC.

As shown in Chapter 1, Deudney, while warning of the dangers of mixing national security and environmental security vocabularies, provides a very telling summary of the mind-shift that is required to address the latter agenda. The differences may be summarised as in Table 6.10.[57]

Table 6.10     National security and environmental security

| National security | Environmental security |
|---|---|
| specific threats | diffuse threats |
| others as enemy | ourselves as enemy |
| intended harm | unintended harm |
| short time-scales | long time-scales |

However, the wave of optimism which greeted the emergence of a post-Cold War Security Council was severely deflated by events surrounding the dissolution of Yugoslavia after 1990. After 1990, the Security Council operated without recourse to the veto for a period of almost exactly four years. The veto was revived by the Russian Federation, in their decision of 11 May 1993, to block the transfer of Cyprus peacekeeping costs to a regular formula after relying since 1962 on *ad hoc* voluntary funding. The previous veto had been that cast by the USA on 31 May 1990, rejecting the creation of a commission of inquiry into the Israeli-occupied territories. A moment's reflection on the agonies of indecision which accompanied the members' dealings with the Bosnian question, 1992–93 shows how much more than the mere absence of the veto is necessary to turn texts into action. It should also be noted that fear of the Russian veto inhibited the adoption of further sanctions against Federal Yugoslavia during the April 1993 Russian referendum campaign.

TV pictures of trench warfare and the deliberate mass-starvation of civilians in Bosnia during 1993 displaced the so called 'spirit of Rio' from most people's consciousness. And yet, as shown in the opening remarks of this book, the 12. 9 million Third World children who die every year from preventable causes represent a loss of human life greatly worse than Bosnia's, in fact a level of human destruction equivalent to repeating the Hiroshima–Nagasaki bombing every three days. This silent holocaust goes unnoticed. No shuttle-diplomacy attends the fate of the 35 000, 'baby-Irma's' who die, daily, unrecorded by the tabloid front page. At the same time, the OECD recession since 1990, and the cheap-energy glut enjoyed since 1986 have combined to undermine the two most obvious incentives to the adoption of sustainable development patterns in the developed countries. The middle of a recession, awash with cheap competitive sources of energy, is not the most propitious time to raise industrial costs, tighten standards of regulation or embark upon unilateral acts of standard-raising and new energy taxes.

That, at least, appears to be the logic in Europe and North America. The Japanese pursuit of lean production, and the shock it will give to the mass-production dinosaurs might prove to be the best stimulus to energy-efficiency, the greening of industrial society and thus some imperfect version of sustainable development that we are likely to be offered within the conventional political and economic possibilities.[58] For the developing countries, the single greatest opportunity to embrace sustainable development lies in reversing the net transfer of wealth from poor to rich that has continued since 1985 as interest payments on debt have outstripped ODA. Massive concessions and debt-forgiveness will generate funds for sustainable development far in excess of any expansion of GEF or ODA. The NATO countries might feel happier about this if they adopted the vocabulary of 'environmental security' to describe and justify this expenditure on debt relief. They could begin to *finance* it by earmarking for this purpose the $50 billion per annum reduction in defence expenditure that has *already* occurred between 1985 and 1992. Finally, who speaks for the common-heritage territories, the 'global commons'? It is time to recognise the damage that has been done by the neglect of the commons during more than a decade of ideological *laissez-faire*. That does not imply any need to revive the corpse of state socialism. Rather it means that an eclectic mix of market-pricing, carbon-taxing and tradeable permits can create incentives to value and so protect the commons when *dirigisme* is inappropriate and verification weak. At the same time, the *appropriate* UN mechanisms of international regulation must be strengthened. There is no clearer guide to this task than *Agenda 21*. What is needed is the political will to transform that document into action. The proper measure of that political will is the readiness of the developed countries to fund the forgiveness of Third World debt, the progressive conversion of defence expenditure and the full funding of ODA and UN reform. Only by these means will both 'sustainable development' and 'environmental security' be elevated from rhetoric to reality.

# Appendix 1
# Declaration of the United Nations Conference on the Human Environment, Part II (The Stockholm Principles, 1972)

## II
### Principles

*States the common conviction that:*

#### Principle 1

Man has the fundamental right to freedom, equality and adequate conditions of life, in an environment of a quality that permits a life of dignity and well-being, and he bears a solemn responsibility to protect and improve the environment for present and future generations. In this respect, policies promoting or perpetuating *apartheid*, racial segregation, discrimination, colonial and other forms of oppression and foreign domination stand condemned and must be eliminated.

#### Principle 2

The natural resources of the earth, including the air, water, land, flora and fauna and especially representative samples of natural ecosystems, must be safeguarded for the benefit of present and future generations through careful planning or management, as appropriate.

#### Principle 3

The capacity of the earth to produce vital renewable resources must be maintained and, wherever practicable, restored or improved.

#### Principle 4

Man has a special responsibility to safeguard and wisely manage the heritage of wildlife and its habitat, which are now gravely imperilled by a combination of adverse factors. Nature conservation, including wildlife, must therefore receive importance in planning for economic development.

#### Principle 5

The non-renewable resources of the earth must be employed in such a way as to guard against the danger of their future exhaustion and to ensure that benefits from such employment are shared by all mankind.

*Principle 6*

The discharge of toxic substances or of other substances and the release of heat, in such quantities or concentrations as to exceed the capacity of the environment to render them harmless, must be halted in order to ensure that serious or irreversible damage is not inflicted upon ecosystems. The just struggle of the peoples of all countries against pollution should be supported.

*Principle 7*

States shall take all possible steps to prevent pollution of the seas by substances that are liable to create hazards to human health, to harm living resources and marine life, to damage amenities or to interfere with other legitimate uses of the sea.

*Principle 8*

Economic and social development is essential for ensuring a favourable living and working environment for man and for creating conditions on earth that are necessary for the improvement of the quality of life.

*Principle 9*

Environmental deficiencies generated by the conditions of under-development and natural disasters pose grave problems and can best be remedied by accelerated development through the transfer of substantial quantities of financial and technological assistance as a supplement to the domestic effort of the developing countries and such timely assistance as may be required.

*Principle 10*

For the developing countries, stability of prices and adequate earnings for primary commodities and raw materials are essential to environmental management since economic factors as well as ecological processes must be taken into account.

*Principle 11*

The environmental policies of all States should enhance and not adversely affect the present or future development potential of developing countries, nor should they hamper the attainment of better living conditions for all, and appropriate steps should be taken by States and international organizations with a view to reaching agreement on meeting the possible national and international economic consequences resulting from the application of environmental measures.

*Principle 12*

Resources should be made available to preserve and improve the environment, taking into account the circumstances and particular requirements of developing countries and any costs which may emanate from their incorporating environmental safeguards into their development planning and the need for making available to them, upon their request, additional international technical and financial assistance for this purpose.

*Principle 13*

In order to achieve a more rational management of resources and thus to improve the environment, States should adopt an integrated and co-ordinated approach to their development planning so as to ensure that development is compatible with the need to protect and improve environment for the benefit of their population.

*Principle 14*

Rational planning constitutes an essential tool for reconciling any conflict between the needs of development and the need to protect and improve the environment.

*Principle 15*

Planning must be applied to human settlements and urbanization with a view to avoiding adverse effects on the environment and obtaining maximum social, economic and environmental benefits for all. In this respect, projects which are designed for colonialist and racist domination must be abandoned.

*Principle 16*

Demographic policies which are without prejudice to basic human rights and which are deemed appropriate by Governments concerned should be applied in those regions where the rate of population growth of excessive population concentrations are likely to have adverse effects on the environment of the human environment and impede development.

*Principle 17*

Appropriate national institutions must be entrusted with the task of planning, managing or controlling the environmental resources of States with a view to enhancing environmental quality.

*Principle 18*

Science and technology, as part of their contribution to economic and social development, must be applied to the identification, avoidance and control of environmental risks and the solution of environmental problems and for the common good of mankind.

*Principle 19*

Education in environmental matters, for the younger generation as well as adults, giving due consideration to the underprivileged, is essential in order to broaden the basis for an enlightened opinion and responsible conduct by individuals, enterprises and communities in protecting and improving the environment in its full human dimension. It is also essential that mass media of communications avoid contributing to the deterioration of the environment, but, on the contrary, disseminate information of an educational nature on the need to protect and improve the environment in order to enable man to develop in every respect.

*Principle 20*

Scientific research and development in the context of environmental problems, both national and multi-national, must be promoted in all countries, especially the developing countries. In this connexion, the free flow of up-to-date scientific information and transfer of experience must be supported and assisted, to facilitate the solution of environmental problems; environmental technologies should be made available to developing countries on terms which would encourage their wide dissemination without constituting an economic burden on the developing countries.

*Principle 21*

States have, in accordance with the Charter of the United Nations and the principles of international law, the sovereign right to exploit their own resources pursuant to their own environmental policies, and the responsibility to ensure that activities within their jurisdiction or control do not cause damage to the environment of other States or of areas beyond the limits of national jurisdiction.

### Principle 22

States shall co-operate to develop further the international law regarding liability and compensation for the victims of pollution and other environmental damage caused by activities within the jurisdiction or control of such States to areas beyond their jurisdiction.

### Principle 23

Without prejudice to such criteria as may be agreed upon by the international community, or to standards which will have to be determined nationally, it will be essential in all cases to consider the systems of values prevailing in each country, and the extent of the applicability of standards which are valid for the most advanced countries but which may be inappropriate and of unwarranted social cost for the developing countries.

### Principle 24

International matters concerning the protection and improvement of the environment should be handled in a co-operative spirit by all countries, big and small, on an equal footing. Co-operation through multilateral or bilateral arrangements or other appropriate means is essential to effectively control, prevent, reduce and eliminate adverse environmental effects resulting from activities conducted in all spheres, in such a way that due account is taken of the sovereignty and interests of all States.

### Principle 25

States shall ensure that international organizations play a co-ordinated, efficient and dynamic role for the protection and improvement of the environment.

### Principle 26

Man and his environment must be spared the effects of nuclear weapons and all other means of mass destruction. States must strive to reach prompt agreement, in the relevant international organs, on the elimination and complete destruction of such weapons.

*21st plenary meeting*
*16 June 1972*

# Appendix 2
# General Assembly Resolution 43/196

*A United Nations conference on environment and development*
Date: 20 December 1988        Meeting; 83
Adopted without a vote        Report; A/43/915/Add. 7

*The General Assembly,*

*Recalling* its resolution 42/186 of 11 December 1987, by which it adopted the Environmental Perspective to the Year 2000 and Beyond as a broad framework to guide national action and international co-operation on policies and programmes aimed at achieving environmentally sound development,

*Recalling also* its resolution 42/187 of 11 December 1987, by which it welcomed the report of the World Commission on Environment and Development,[64]

*Noting* that the United Nations Conference on the Human Environment, convened in 1972 in accordance with Assembly resolution 2398 (XXIII) of 3 December 1968, recommended that the Assembly convene a second United Nations conference on the subject,[65]

*Believing* it highly desirable that a United Nations conference on environment and development be convened no later than 1992,

*Aware* that serious environmental problems are arising in all countries and that these problems must be progressively addressed through preventive measures at their source,

*Emphasizing* the common goal of all countries to strengthen international co-operation for the promotion of growth and development world-wide and recognizing that, in view of the global character of major environmental problems, there is a common interest of all countries in pursuing policies aimed at achieving a sustainable and environmentally sound development within a sound ecological balance,

*Noting* that the critical objectives for environment and development policies that follow from the need for sustainable and environmentally sound development must include creating a healthy, clean and safe environment, reviving growth and improving its quality, remedying the problems of poverty and the satisfaction of

---

64. A/42/427, annex.
65. *Report of the United Nations Conference on the Human Environment, Stockholm, 5–16 June 1972* (United Nations publication, Sales No. E. 73. II. A. 14 and Corr.1), chap. IV, resolution 4 (I).

human needs through raising the standard of living and the quality of life, addressing the issues of population and of conserving and enhancing the resource base, reorienting technology and managing risk and merging environment and economics in decision-making,

*Aware* that a supportive international economic environment that would result in sustained economic growth and development in all countries, particularly in developing countries, is of major importance for sound management of the environment,

*Stressing* the importance for all countries to take effective measures for the protection, restoration and enhancement of the environment in accordance, *inter alia*, with their respective capabilities, while, at the same time, acknowledging the efforts being made in all countries in this regard, including international co-operation between developed and developing countries,

*Noting* the fact that the largest part of the current emission of pollutants into the environment, including toxic and hazardous wastes, originates in developed countries, and therefore recognizing that those countries have the main responsibility for combating such pollution,

*Reaffirming* the need for additional financial resources from the international community effectively to support developing countries in identifying, analysing, monitoring, preventing and managing environment problems in accordance with their national development plans, priorities and objectives,

*Reaffirming* the need for developed countries and the appropriate organs and organizations of the United Nations system to strengthen technical co-operation with the developing countries to enable them to develop and strengthen their capacity for identifying, analysing, monitoring, preventing and managing environmental problems in accordance with their national development plans, priorities and objectives,

*Recognizing* the importance of international co-operation in the research and development of environmentally sound technology and the need for an international exchange of experience and knowledge as well as the promotion of the transfer of technology for the protection and enhancement of the environment, especially in developing countries, in accordance with national laws, regulations and policies,

*Reaffirming* the need for support by the international community in playing a catalytic role in technical co-operation among developing countries in the field of the environment, and inviting the appropriate organs and organizations of the United Nations system to co-operate, at the request of the parties concerned, in the promotion and strengthening of such co-operation,

*Aware* that threats to the environment often have a transboundary impact and that their urgent nature requires strengthened international co-operative action, *inter alia*, by assessing and providing early warning to the world community on serious environmental threats within the framework of Earthwatch,

*Taking note with appreciation* of the progress report of the Secretary-General on the implementation of General Assembly resolution 42/187,[66]

*Noting further* that, by its resolution 42/187, the General Assembly invited Governments, in co-operation with the regional commissions and the United Nations Environment programme and, as appropriate, intergovernmental organizations, to support and engage in follow-up activities, such as conferences, at the national, regional and global levels,

*Noting* the importance of exploring ways and means of how best to promote sustainable and environmentally sound development in all countries, taking into account General Assembly resolutions 42/186 and 42/187,

*Considering also* in this context that the conference could, *inter alia*:

(*a*) Review trends in policies and action taken by all countries and international organizations to protect and enhance the environment and to examine how environmental concerns have been incorporated in economic and social policies and planning since the United Nations Conference on the human Environment in 1972;

(*b*) Assess major environmental problems, risks and opportunities associated with economic activities in all countries;

(*c*) Make recommendations for further strengthened international co-operative action within a set of priorities to be established by the conference; define the research and development effort required to implement such recommendations; and indicate financial requirements for their implementation, together with a definition of possible sources for such financing;

1. *Decides* to consider at its forty-fourth session the question of the convening of a United Nations conference on the subject of the present resolution no later than 1992, with a view to taking an appropriate decision at that session on the exact scope, title, venue and date of such a conference and on the modalities and financial implications of holding the conference;

2. *Requests* the Secretary-General, with the assistance of the Executive Director of the United Nations Environment Programme, urgently to obtain the views of Governments on:

(*a*) The objectives, content, title and scope of the conference;
(*b*) Appropriate ways of preparing for the conference;
(*c*) A suitable time and place and other modalities for the conference;

and to submit those views to the General Assembly at its forty-fourth session, through the Economic and Social Council, and to make them available to the Governing Council of the United Nations Environment Programme at its fifteenth session;

3. *Also requests* the Secretary-General, with the assistance of the Executive Director, to obtain the views of appropriate organs, organizations and programmes

---

66. A/43/353-E/1988/71.

of the United Nations system and relevant intergovernmental and non-governmental organizations on the objectives, content and scope of the conference, and to submit those views to the Assembly at its forty-fourth session, through the Economic and Social Council, and to make them available to the Governing Council at its fifteenth session;

4. *Further requests* the Secretary-General, with the assistance of the Executive Director, to prepare a statement of the financial implications of preparing and convening the conference, and to submit that statement to the General Assembly at its forty-fourth session, through the Economic and Social Council, and to make it available to the Governing Council at its fifteenth session;

5. *Invites* the Governing Council to consider the documents referred to in paragraphs 2 to 4 above and, on the basis of that consideration, to submit to the General Assembly at its forty-fourth session, through the Economic and Social Council, its views on the matters referred to in the present resolution, in particular its views on the objectives, content and scope of the conference.

# Appendix 3
# Rio Declaration on Environment and Development

*Having met* at Rio de Janeiro from 3 to 14 June 1992,

*Reaffirming* the Declaration of the United Nations Conference on the Human Environment, adopted at Stockholm on 16 June 1972,[1] and seeking to build upon it,

*With the goal of* establishing a new and equitable global partnership through the creation of new levels of cooperation among States, key sectors of societies and people,

*Working towards* international agreements which respect the interests of all and protect the integrity of the global environmental and developmental system,

*Recognizing* the integral and interdependent nature of the Earth, our home,

*Proclaims that:*

PRINCIPLE 1
Human beings are at the centre of concerns for sustainable development. They are entitled to a healthy and productive life in harmony with nature.

PRINCIPLE 2
States have, in accordance with the Charter of the United Nations and the principles of international law, the sovereign right to exploit their own resources pursuant to their own environmental and developmental policies, and the responsibility to ensure that activities within their jurisdiction or control do not cause damage to the environment of other States or of areas beyond the limits of national jurisdiction.

PRINCIPLE 3
The right to development must be fulfilled so as to equitably meet developmental and environmental needs of present and future generations.

PRINCIPLE 4
In order to achieve sustainable development, environmental protection shall constitute an integral part of the development process and cannot be considered in isolation from it.

---

1. See Appendix 1.

PRINCIPLE 5

All States and all people shall cooperate in the essential task of eradicating poverty as an indispensable requirement for sustainable development, in order to decrease the disparities in standards of living and better meet the needs of the majority of the people of the world.

PRINCIPLE 6

The special situation and needs of developing countries, particularly the least developed and those most environmentally vulnerable, shall be given special priority. International actions in the field of environment and development should also address the interests and needs of all countries.

PRINCIPLE 7

States shall cooperate in a spirit of global partnership to conserve, protect and restore the health and integrity of the Earth's ecosystem. In view of the different contributions to global environmental degradation, States have common but differentiated responsibilities. The developed countries acknowledge the responsibility that they bear in the international pursuit of sustainable development in view of the pressures their societies place on the global environment and of the technologies and financial resources they command.

PRINCIPLE 8

To achieve sustainable development and a higher quality of life for all people, States should reduce and eliminate unsustainable patterns of production and consumption and promote appropriate demographic policies.

PRINCIPLE 9

States should cooperate to strengthen endogenous capacity-building for sustainable development by improving scientific understanding through exchanges of scientific and technological knowledge, and by enhancing the development, adaptation, diffusion and transfer of technologies, including new and innovative technologies.

PRINCIPLE 10

Environmental issues are best handled with the participation of all concerned citizens, at the relevant level. At the national level, each individual shall have appropriate access to information concerning the environment that is held by public authorities, including information on hazardous materials and activities in their communities, and the opportunity to participate in decision-making processes. States shall facilitate and encourage public awareness and participation by making information widely available. Effective access to judicial and administrative proceedings, including redress and remedy, shall be provided.

PRINCIPLE 11

States shall enact effective environmental legislation. Environmental standards, management objectives and priorities should reflect the environmental and developmental context to which they apply. Standards applied by some countries may be inappropriate and of unwarranted economic and social cost to other countries, in particular developing countries.

PRINCIPLE 12

States should cooperate to promote a supportive and open international economic system that would lead to economic growth and sustainable development in all countries, to better address the problems of environmental degradation. Trade policy measures for environmental purposes should not constitute a means of arbitrary or unjustifiable discrimination or a disguised restriction on international trade. Unilateral actions to deal with environmental challenges outside the jurisdiction of the importing country should be avoided. Environmental measures addressing transboundary or global environmental problems should, as far as possible, be based on an international consensus.

PRINCIPLE 13

States shall develop national law regarding liability and compensation for the victims of pollution and other environmental damage. States shall also cooperate in an expeditious and more determined manner to develop further international law regarding liability and compensation for adverse effects of environmental damage caused by activities within their jurisdiction or control to areas beyond their jurisdiction.

PRINCIPLE 14

States should effectively cooperate to discourage or prevent the relocation and transfer to other States of any activities and substances that cause severe environmental degradation or are found to be harmful to human health.

PRINCIPLE 15

In order to protect the environment, the precautionary approach shall be widely applied by States according to their capabilities. Where there are threats of serious or irreversible damage, lack of full scientific certainty shall not be used as a reason for postponing cost-effective measures to prevent environmental degradation.

PRINCIPLE 16

National authorities should endeavour to promote the internalization of environmental costs and the use of economic instruments, taking into account the approach that the polluter should, in principle, bear the cost of pollution, with due regard to the public interest and without distorting international trade and investment.

PRINCIPLE 17

Environmental impact assessment, as a national instrument, shall be undertaken for proposed activities that are likely to have a significant adverse impact on the environment and are subject to a decision of a competent national authority.

PRINCIPLE 18

States shall immediately notify other States of any natural disasters or other emergencies that are likely to produce sudden harmful effects on the environment of those States. Every effort shall be made by the international community to help States so afflicted.

PRINCIPLE 19

States shall provide prior and timely notification and relevant information to potentially affected States on activities that may have a significant adverse transboundary environmental effect and shall consult with those States at an early stage and in good faith.

PRINCIPLE 20

Women have a vital role in environmental management and development. Their full participation is therefore essential to achieve sustainable development.

PRINCIPLE 21

The creativity, ideals and courage of the youth of the world should be mobilized to forge a global partnership in order to achieve sustainable development and ensure a better future for all.

PRINCIPLE 22

Indigenous people and their communities and other local communities have a vital role in environmental management and development because of their knowledge and traditional practices. States should recognize and duly support their identity, culture and interests and enable their effective participation in the achievement of sustainable development.

PRINCIPLE 23

The environment and natural resources of people under oppression, domination and occupation shall be protected.

PRINCIPLE 24

Warfare is inherently destructive of sustainable development. States shall therefore respect international law providing protection for the environment in times of armed conflict and cooperate in its further development, as necessary.

PRINCIPLE 25

Peace, development and environmental protection are interdependent and indivisible.

PRINCIPLE 26

States shall resolve all their environmental disputes peacefully and by appropriate means in accordance with the Charter of the United Nations.

PRINCIPLE 27

States and people shall cooperate in good faith and in a spirit of partnership in the fulfilment of the principles embodied in this Declaration and in the further development of international law in the field of sustainable development.

# Notes

## 1 TWO HIROSHIMAS EVERY WEEK

1. The figure of 12.9 million annual child deaths is supplied by World Resources Institute, *World Resources 1992–93*, (UNEP-UNDEP-Oxford University Press, 1992), p. 82. The figure of Hiroshima and Nagasaki fatalities combines the death-toll recorded up to four months after the detonations: John May, *The Greenpeace Book of the Nuclear Age* (London: Gollancz, 1989), pp. 74–6.
2. World Resources Institute, op. cit., p. 167.
3. *Agenda 21; The United Nations Programme of Action from Rio*, UN, 1993, p. 175.
4. Ibid., p. 177.
5. Susan George, *The Debt Boomerang* (Westwiew, 1992), pp. xvi–xvii.
6. M. Grubb *et al.*, *The Earth Summit Agreements*, Royal Institute of International Affairs, 1993, pp. 174–7.
7. 'The Global Environmental Facility', *Our Planet*, Vol. 3 (1991), pp. 10–13.
8. As of 6 October 1993, the UN regular budget was owed $536 million and peacekeeping accounts were owed $1200 million, a combined total of $1.73 billion. *Washington Weekly Report*, XIX-30, 7 October 1993.
9. *Agenda 21*, op. cit., Chapter 33.18, p. 251.
10. The definition of sustainable development is from the so-called Brundtland Report, properly styled, the World Commission on Environment and Development, *Our Common Future* (Oxford: Oxford University Press, 1987) p. 8. Over forty-five various definitions of sustainable development are collected in D. Pearce, A. Markandya and E. Barbier, *Blueprint for a Green Economy* (London: Earthscan, 1989), Annex, pp. 173–5.
11. A. Westing (ed.), *Global Resources and International Conflicts* (SIPRI-Oxford University Press, 1986), pp. 204–9 tabulates the resource-control component in a number of twentieth-century conflicts such as the Chaco War, the Congo–Katanga secession, and the Nigerian civil war.
12. See W. D. Nordhaus, 'To Slow or Not to Slow; The Economics of the Greenhouse Effect', *Economic Journal*, Vol. 101 (1991), pp. 920–37.
13. See R. Keohane and J. Nye, *Power and Interdependence* (New York: Little, Brown, 1977). Also S. Krasner (ed.), International Regimes (Ithaca: Cornell University Press, 1983).
14. See, for example, I. Roxborough, *Theories of Underdevelopment* (London: Macmillan, 1979).
15. C. Thomas, *The Environment in International Relations* (Royal Institute of International Affairs, 1992), Chapter 4.
16. J. Herz, *International Politics in the Atomic Age* (Columbia, 1959), pp. 96–108.
17. See, for example, S. Hassan, *Environmental Issues and Security in South Asia*, Adelphi Paper No. 262, International Institute of Strategic Studies, 1991.

18. See Gil Loescher, *Refugee Movements and International Security*, Adelphi Paper No. 268, International Institute of Strategic Studies, 1991.
19. A wide range of issues is addressed in A. Westing (ed.), *Global Resources and International Conflict* (SIPRI–Oxford University Press, 1986), also, *Comprehensive Security for the Baltic; an Environmental Approach* (Sage, 1989), and *Environmental Warfare* (Taylor and Francis, 1984).
20. See A. Westing (ed.) *Explosive Remnants of War; Mitigating the Environmental Effects* (SIPRI–Taylor and Francis, 1985), also *Herbicides in War* (SIPRI–Taylor and Francis, 1984), and *Environmental Warfare* (SIPRI–Taylor and Francis, 1984).
21. A. Anderson, 'The Environmental Aftermath of the Kuwait–Iraq Conflict', *Our Planet*, Vol. 3, 1991, p. 4.
22. World Commission on Environment and Development, op. cit., p. 303.
23. D. Deudney, 'The Case Against Linking Environmental Degradation and National Security', *Millennium*, Vol. 19, 1990, p. 473.
24. C. Thomas, op. cit., pp. 117–19.
25. S. Evteev, R. A. Perelet and V. P. Voronin, 'Ecological Security and Sustainable Development', in J. Renninger (ed.), *The Future of the United Nations in an Interdependent World* (Martinus Nijhoff–UNITAR, 1989), pp. 162–71.
26. David Kay and Eugene B. Skolnikoff, *World Eco-Crisis, International Organizations in Response* (Wisconsin, 1972). R. Boardman, *International Organization and the Conservation of Nature* (London: Macmillan, 1981). D. Kay and H. K. Jacobson, (eds), *Environmental Protection, The International Dimension* (Allanheld, Osmun, 1983).
27. See, The Brandt Commission, *Common Crisis: North South Cooperation for World Recovery* (London: Pan, 1983). Gerald O. Barney (Director), *The Global 2000 Report to the President* (Harmondsworth: Penguin, 1982). World Commission on Environment and Development, op. cit.
28. R. Keohane and J. Nye, *Power and Interdependence* (New York: Little, Brown, 1973).
29. The functionalist thesis is most widely attributed to David Mitrany; see *The Functional Theory of Politics* (Martin Robertson, 1975), also A. J. Groom and P. Taylor (eds), *Functionalism* (London University Press, 1975), also M. F. Imber, *The USA, ILO, UNESCO and IAEA* (London: Macmillan, 1989).
30. On the prisoner's dilemma, see R. Axelrod, *The Evolution of Cooperation* (Harmonsworth: Penguin, 1990).
31. Adapted from H. K. Jacobson, *Networks of Interdependence* (New York: Alfred Knopf, 1979).
32. See M. F. Imber, op. cit., pp. 28–41.
33. See, for example, J. Galtung, 'A structural theory of imperialism', *Journal of Peace Research*, Vol. 13, 1971, pp. 81–94. Also I. Wallerstein, 'Dependency ... world systems. ...The rise and future demise of the world capitalist system: concepts for comparative analysis', *Comparative Studies in Society and History*, Vol. 16, 1974, pp. 387–415.
34. Roxborough, op. cit., pp. 28–32.
35. R. Keohane, *After Hegemony* (Princeton, 1984), also *International Institutions and State Power* (Westview, 1989).

36. See J. McCormick, *The International Environmental Movement; Reclaiming Paradise* (Belhaven, 1989). Also Douglas Williams, *The Specialised Agencies and the United Nations* (Hurst, 1986).
37. See D. R. Marples, *Chernobyl and Nuclear Power in the USSR* (London: Macmillan, 1987).
38. C. Hermann, 'International crisis as a situational variable', in J. Rosenau (ed.), *International Politics and Foreign Policy* (Free Press, 1969), pp. 409–21.
39. S. Saetevik, *Environmental Cooperation between North Sea States* (Belhaven, 1988), p. 10.
40. Nordhaus, op. cit.
41. K. Galbraith, *The Culture of Contentment* (London: Sinclair-Stevenson, 1992). See also A. Hirschman, *Exit, Voice and Loyalty* (Harvard, 1970), pp. 44–5, and M. Imber, op. cit., pp. 139–40.
42. *Agenda 21*, op. cit., Chapter 33.18, p. 251. See also M. Grubb, et al., op. cit., Appendix, pp. 170–3.
43. Young, op. cit., passim, especially Chapter 3.
44. A. Dobson, *Green Political Thought* (London: Unwin Hyman, 1990), p. 13.
45. The common heritage of mankind is a concept of property recognised by the adoption of UN General Assembly resolution 2749 (XXV) of 17 December 1970. The concept was further developed within the United Nations Conference of the Law of the Sea, 1977–82. It is enshrined in the provisions of the United Nations Convention of 1982, and applies to the seabed beyond the limits of national jurisdiction. As the name implies, common heritage resources are neither the sovereign possession of one state, nor *res nullius*, open to all, like the high seas. The common heritage of mankind requires some collective form of administration. In the case of the law of the sea this will be the International Seabed Authority of the UN. Due to insufficient ratification these provisions of the 1982 Convention are not yet in force. See *United Nations Convention on the Law of the Sea* (United Nations, 1983), preamble, also Article 136.
46. See Pearce, et. al., op. cit., pp. 51–81.
47. A. Dobson, op. cit., discusses ecologism, socialism and eco-feminism, pp. 171–204. J. Young, op. cit., discusses the Gaia hypothesis, pp. 118–36. Also J. Saurin, 'Global environmental degradation, modernity and environmental knowledge', *Environmental Politics*, Vol. 2, 1993.

## 2 DEBT, POVERTY AND ENVIRONMENT

1. Morris Miller, *Debt and the Environment; Converging Crises* (United Nations, 1991), p. 11.
2. Susan George, *The Debt Boomerang* (Westview, 1992), pp. xv–xvi. Miller, op. cit., pp. 14–15; figures are based on longer-term debt.
3. George, op. cit., pp. xv–xvi.
4. Miller, op. cit., p. 11.
5. World Resources Institute, *World Resources 1992–93* (UNEP–UNDP–Oxford University Press, 1992), p. 82.

6. Quotation from Paul Vallely, *Bad Samaritans* (London: Hodder and Stoughton, 1990), p. 190. See also Richard Jolly, 'The Human Dimensions of International Debt', in *Growing Out of Debt*, A. Hewitt and B. Wells (eds) for the All Party Parliamentary Group on Overseas Development, Overseas Development Institute, 1989, pp. 51–2.

7. World Resources Institute, op. cit., pp. 82–3.

8. Ibid., pp. 84–5.

9. The release of 40 tonnes of methyl isocyanate from the Union Carbide plant on the night of 2–3 December 1984 caused the deaths of 2500 local residents, and injury to 30 000 more. See David Weir, *The Bhopal Syndrome* (Earthscan, 1987).

10. Ibid., pp. 99–101.

11. The final declaration of the Stockholm Conference on the Human Environment, included Principle 21, 'States have, in accordance with international law, the sovereign right to exploit their own natural resources pursuant to their own environmental policies, and the responsibility to ensure that activities within their jurisdiction or control do not cause damage to the environment of other states or of areas beyond the limits of national jurisdiction'. *Report of the United Nations Conference on the Human Environment*, United Nations, A/CONF. 48/14/Rev. 1.

12. See *Conservation and Development of Forests*, United Nations, A/CONF. 151/PC/65 11 July 1991, p. 4. Indonesia objected to the provision of specific language on this question, claiming that equal civil-rights provisions in that country made special language unnecessary.

13. *Agenda 21, The United Nations Programme of Action from Rio*, UN, 1993, Chapter 26, pp. 227–9.

14. George, op. cit., p. 25.

15. For two succinct contributions by Norman Myers, see in particular 'Tropical Forests', in J. Leggett (ed.), *Global Warming, The Greenpeace Report* (Oxford: Oxford University Press, 1990), pp. 372–99, and 'The Anatomy of Environmental Action; The Case of Tropical Deforestation', in Andrew Hurrell and Benedict Kingsbury (eds), *The International Politics of the Environment* (Oxford: Oxford University Press, 1992), pp. 430–54.

16. George, op. cit., p. 10.

17. Myers, op. cit., 1992, pp. 445–7.

18. George, op. cit., pp. 12–13.

19. George, op. cit., p. 23

20. See M. Imber, *The USA, ILO, UNESCO and IAEA, Withdrawal and Politicization in the Specialized Agencies* (London: Macmillan, 1989), pp. 122–36.

21. J. Williamson, *The Open Economy and the World Economy* (Basic, 1983), p. 307.

22. Miller, op. cit., p. 52.

23. Williamson, op. cit., p. 295.

24. Ibid., pp. 295–6.

25. Miller, op. cit., p. 56.

26. Ibid., p. 26.

27. Ibid., pp. 54–5.

28. Ibid., pp. 64–5.

29. Ibid., p. 66.
30. J. Williamson, op. cit., p. 111.
31. J. T. Rourke, *International Politics on the World Stage* (Dushkin Publishing, 2nd edn, 1989), p. 380.
32. World Resources Institute, op. cit., p. 42.
33. George, op. cit., passim.
34. Miller, op. cit., pp. 294–333.
35. Both Miller and Vallely discuss a wide range of plans for debt relief and debt swaps. See Miller, op. cit., pp. 118–24, 145–58 and p. 185. Also Vallely, op. cit., pp. 303–6.
36. See Miller, op. cit., pp. 297–8.
37. Ibid., p. 122.
38. Ibid., p. 126.
39. Ibid., p. 128.
40. J. Williamson, op. cit., p. 365.
41. Miller, op. cit., p. 114.

## 3 THE GLOBAL COMMONS

1. Quoted by Clyde Sanger, *Ordering the Oceans* (Zed, 1986), p. 158.
2. Sanger, op. cit., p. 158.
3. C. Pinto, 'Towards a regime governing international public property', in Antony J. Dolman, (ed.), *Global Planning and Resources Management* (London: Pergamon, 1980), p. 208.
4. Pinto, op. cit., p. 209.
5. See F. Berkes, (ed.), *Common Property Resources, Ecology and Community Based Sustainable Development* (Belhaven, 1989).
6. See G. Hardin, 'The Tragedy of the Commons', *Science*, Vol. 162, 1968, pp. 1243–8. Also G. Hardin and J. Baden, (eds), *Managing the Commons* (W. H. Freeman, 1978).
7. Susan J. Buck (Cox), 'Multi jurisdictional resources: testing a typology of problem solving', in Berkes, op. cit., pp. 127–8.
8. Creeping territoriality is coined and discussed by M. Imber, 'International Institutions and the Common Heritage of Mankind, Sea, Space and Polar Regions', in P. Taylor and A. J. Groom (eds), *International Institutions at Work* (Pinter, 1988), p. 150.
9. See Sanger, op. cit., pp. 70–89.
10. *United Nations Convention on the Law of the Sea*, 1982, Part XI, especially Articles 150–1.
11. Ibid., Articles 117–20, describe the obligations of states to practise conservation measures on the high-sea fish stocks. Annexe I, identifies seventeen species of highly migratory fish and cetaceans, for which states are obliged to observe conservation practices even within their exclusive economic zones, in effect creating rights and duties on states to conserve shared resources.
12. *Agenda 21*, op. cit., Chapter 17. 44.

13. Sanger, op. cit., pp. 48–52. For fuller discussion of the US decision not to sign, see A. Hollick, *US Foreign Policy and the Law of the Sea* (Princeton, 1981), and B. Oxman, D. Carson and C. Buderi, *The Law of the Sea; a US policy dilemma* (California Institute of Contemporary Studies, 1983).

14. The most generous extension of continental-shelf rights was allowed by a complex formula described in Article 76, paras 5, 6.

15. The role of the ISA and the revenue-sharing arrangements are contained in Part XI, especially Articles, 136, 137, 140, 150–63.

16. D. R. Denman, *Markets Under the Sea?* (Institute of Economic Affairs, 1984), makes the case for private property titles to the deep-seabed.

17. See C. Sanger, op. cit., pp. 48–55. See also the *Republican Party Platform*, 1984, p. 10, 'The President [Reagan] decisively rejected the UN Convention on the Law of the Sea and embarked instead on a dynamic national oceans policy, animated by our traditional commitment to freedom of the seas. That pattern will be followed with respect to UN meddling in Antarctica and Outer Space.'

18. Safire, writing in *The New York Times*, is quoted by Sanger, op. cit., p. 51.

19. See United Nations, *Convention on the Law of the Sea*, 1982, Article 161.

20. For a full description of both the negotiation and terms of the regime, see Owen Greene, 'Ozone Depletion: Implementing and Strengthening the Montreal Protocol', in J. B. Poole and R. Guthrie (eds), *Verification Report 1992* (VERTIC, 1992), pp. 265–74.

21. Ibid., p. 269.

22. See A. Westing, (ed.) *Environmental Warfare* (SIPRI–Taylor and Francis, 1984), contains three chapters on ENMOD by Erno Meszaros, Jozef Goldblat and Allan S. Krass respectively.

23. Subrata Roy Chowdhury, 'Permanent sovereignty over natural resources', in K. Hossain and S. Roy Chowdhury, *Permanent Sovereignty over Natural Resources in International Law* (Pinter, 1984), p. 1.

24. Rachel McCleary, 'The International Community's Claim to Rights in Brazilian Amazonia', *Political Studies*, Vol. XXXIX, December 1991, pp. 691–707.

25. Ibid., pp. 691–2.

26. Ibid., p. 706.

27. *Statement by His Excellency Dr Mahatir Mohamed, Prime Minister of Malaysia*, at UNCED, Rio, 13 June 1992. Source: Malaysian Mission to the United Nations, New York.

28. See World Commission on Environment and Development, *Our Common Future* (Oxford, 1986), p. 276. For full discussion of the common-heritage qualities of radio frequencies, see John Vogler, 'Regimes and the Global Commons: Space, Atmosphere and Oceans', in A. McGrew and P. Lewis, *Global Politics* (Polity Press, 1992), pp. 118–37.

29. The most extensive discussion of tradeable permits for carbon-dioxide emissions is contained in M. Grubb, 'The greenhouse effect: negotiating targets', *International Affairs*, Vol. 66, 1990, pp. 67–89.

30. *The Independent*, 10 October 1992.

31. *Report of the United Nations Conference on the Human Environment* (UN, 1972), p. 5.

32. *Agenda 21*, op. cit., Chapter 34, p. 252.

## 4   THE UNEP ROLE

1. John McCormick, *The International Environmental Movement; Reclaiming Paradise* (Belhaven, 1989), p. 91.
2. Ibid., p. 93.
3. For extended commentary and texts, see *United Nations Yearbook*, 1971, pp. 307–13.
4. McCormick, op. cit., pp. 93–4.
5. The full text of the 26 Principles may be found in *Report of the United Nations Conference on the Human Environment*, A/CONF. 48/14/Rev. 1, 1973, pp. 4–5 and is reprintedd as Appendix 1.
6. *UNEP Profile*, UNEP, 1990, p. 32.
7. The delicacy and politicking surrounding demographic issues with the UN are well described by P. Taylor, 'Population: Coming to Terms with People', in P. Taylor and A. J. Groom (eds), *Global Issues in the United Nations' Framework* (London: Macmillan, 1989), pp. 148–76.
8. McCormick, op. cit., p. 103.
9. See 'Trail Smelter Arbitration', *American Journal of International Law*, 1939, p. 182. Also discussed by I. Detter de Lupis, 'The Human Environment' in P. Taylor and A. J. Groom, (eds), op. cit., p. 221.
10. The reference to a small secretariat is taken from Resolution 2997 (XXVII), Section II, para. 1. This Resolution, adopted in December 1972, replicated a resolution adopted in the 17th plenary session at Stockholm on 15 June of that year.
11. All references to United Nations General Assembly Resolution 2997, (XXVII), Section I, paras 1–3, 15 December 1972. Adopted by 116 votes to 0, with 10 abstentions.
12. UN, General Assembly Resolution 2997 (XXVII), Section II, para. 2, (a)–(e).
13. Ibid., Section II, para. 2 (f)–(g).
14. UNEP GC. 15/9/Add. 1, 30 January 1989, 89 pp.
15. See *Our Changing Planet: The FY 1991 US Global Change Research Programme*, Office of Science Technology Policy, Federal Coordinating Council for Science, Engineering and Technology, Executive Office of the President, Washington, DC.
16. United States General Accounting Office, *United Nations, U.S. Participation in the Environment Program*, GAO/NSIAD-89-142, Washington, DC, 1989, pp. 12–13.
17. Ibid., p. 24.
18. UNEP, *Evaluation Report 1988*, UNEP /RG/88/2, Nairobi, 1988, pp. 56–58.
19. UNEP, *Evaluation Report 1989*, Nairobi, 1989, p. 16.
20. Ibid., p 16.
21. The Department of the Environment is the lead agency for UK policy in UNEP. The Secretary of State is answerable for Parliamentary questions. The Department pays the UK subscription to the Environment Fund, and coordinates the Whitehall machinery on UNEP questions. The FCO is part of the delegation; the ODA is also prominent on aid questions. The Department of Energy has the technical lead on $CO_2$ questions; the DTI has responsibility on technology transfer and emissions; MAF has interest in adaptation

to climate-change. The Treasury is the Treasury. Cooperation in Whitehall is *ad hoc*. There is no US-style system of inter-agency review, or standing committee, although reference has been made to an Interdepartmental Liaison Group meeting every six weeks. Both the House of Commons and House of Lords operate committees on environmental affairs.

## 5 TWO CHEERS FOR RIO, 1992

1. For an excellent survey of the past record, see Ingrid Detter de Lupis, 'The Human Environment: Stockholm and Its Follow Up', in P. Taylor and A. J. Groom, *Global Issues in the UN Framework* (London: Macmillan, 1989), pp. 205–25.

2. M. Grubb, M. Koch, A. Munson, F. Sullivan, K. Thompson, *The Earth Summit Agreements, A Guide and an Assessment* (Royal Institute of International Affairs, 1993).

3. Caroline Thomas (ed.), *Rio: Unravelling the Consequences* (London: Frank Cass, forthcoming 1994). Thomas's preview of the UNCED agenda is contained in *The Environment in International Relations* (Royal Institute of International Affairs, 1992).

4. A. Rogers, *The Earth Summit, A planetary reckoning* (Global View Press, 1993), contains a full appreciation of the non-governmental dimension of the UNCED.

5. Peter H. Sands, *Lessons Learned in Global Governance* (Washington, DC: World Resources Institute, 1990), p. 14.

6. See, M. Imber, 'Environmental security; a task for the UN system', *Review of International Studies*, Vol. 17, 1991, pp. 201–12, especially, pp. 211–12. Also, Auslan Cramb, 'Environment conference will fail, says academic', *The Scotsman*, 7 January 1992.

7. UN General Assembly resolutions 38/161 of 19 December 1983 and 42/186 of 11 December 1987. Published by UNEP as *Environmental Perspective to the Year 2000* (Nairobi: UNEP, 1988).

8. Available as UNEP document GCSSI/&/Add1, 1988.

9. UN General Assembly, A/CONF. 151/PC/6, 27 June 1990, *Overview of system-wide activities relevant to General Assembly resolution 44/228*, 48 pp.

10. Speech by Dato Seri Dr Mahatir Mohammed, Prime Minister of Malaysia, to *2nd Ministerial Conference of Developing Countries on Environment and Development*, Kuala Lumpur, 27 April 1992. Source: Permanent Mission of Malaysia to the United Nations, New York.

11. 153 states signed the Framework Convention on Climate Change at Rio. Six months later, in December 1992, the following countries had *not* signed the FCCC; Albania, Bosnia and Hercegovina, Brunei,* Cambodia, Czechoslovakia, Dominica, Equatorial Guinea, Georgia, Grenada, the Holy See, Iran*, Iraq*, Kuwait *, Kyrgystan, Laos, Malaysia, Panama, Qatar*, St Lucia, St Vincent, Saudi Arabia*, Sierra Leone, Somalia, South Africa, Syria, Tajikistan, Tonga, Turkey, Turkmenistan, UAE*, Uzbekistan (* denotes OPEC member). See Grubb, *et. al.*, op. cit., 1993, p. 15.

12. Ibid., p. 15.

13. The document appeared as (UNCED A/CONF. 151/6/Rev1), later reproduced in *Agenda 21, The United Nations Programme of Action From Rio* (UN, 1993), pp. 289–94.
14. Francis Sullivan, 'Forest Principles', in M. Grubb *et al.*, op. cit., p. 161.
15. The Rio Declaration is also included in *Agenda 21*, op. cit., pp. 9–11.
16. See Adam Rogers, op. cit., p. 193.
17. K. Thompson, 'The Rio Declaration', in M. Grubb *et al.*, op. cit., p. 91.
18. *Agenda 21*, op. cit., Chapter 33. 18, pp. 251.
19. Ibid., Chapter 33.10, p. 250.
20. Ibid., Chapter 33.18, p. 251.
21. World Commission on Environment and Development, *Our Common Future* (Oxford, 1987), p. 303.
22. United Nations, *The Global Partnership* (United Nations, April 1992), p. 17.
23. M. Grubb, *et al.*, op. cit., p. 174.
24. World Commission on Environment and Development, op. cit., p. 318.
25. Ibid., pp. 318–19
26. This statement is extracted from a Prepcom document prepared by the conference secretariat in advance of the Geneva sessions of August 1991, nine months prior to Rio. See UN, A/CONF.151/PC/80, para. 11.
27. Roddick, *UNCED, its 'Stakeholders' and the Post-UNCED Process*, Mimeo, University of Glasgow, Institute of Latin American Studies, May 1993, pp. 11–12.
28. Ibid., pp. 13–15.
29. Rogers, op. cit., p. 228.
30. ECOSOC comprises 54 elected members drawn from the General Assembly. Eighteen join and leave in rotation each year after a three-year term. See *Basic Facts About the United Nations* (New York: UN, 1987), p. 9.
31. The following list is adapted from that produced by Tom Bigg of the UK–UNA Sustainable Development Unit in March 1993. See also Kathy Sessions, *Washington Weekly Report*, XIX–6, 5 March 1993.
32. N. Myers, *The Gaia Atlas of Future Worlds* (Robertson McCarta, 1990), p. 82.
33. *Agenda 21*, op. cit., Chapter 38.2, p. 274.
34. Ibid., Chapter 38.16, pp. 276–7.
35. Data is complicated by the UN organs operating a biennium funding system, and by the existence of extensive trust funds and reserves as well as straight cash-donations to the two programmes. See General Accounting Office, op. cit., p. 9 and also General Accounting Office, *United Nations; US Participation in the UN Development Programme*, February 1990, p. 29.
36. Rogers, op. cit., passim.
37. P. Doran, 'The Earth Summit, (UNCED): Ecology as Spectacle', *Paradigms*, Vol. 7, 1993, pp. 55–66.
38. Ibid., p. 57.
39. Rogers, op. cit., p. 233.

## 6 BEYOND UNCED: REVENUES AND REFORMS

1. For a balanced view of the UN's problems, see D. Williams, *The Specialized Agencies and the United Nations; The System in Crisis* (Hurst and Co., 1987), especially Chapter V. Also J. Harrod and N. Schrijver (eds), *The UN Under Attack* (Aldershot: Gower, 1988). See also the compendious end-of-the-decade survey contained in M. Karns and K. Mingst (eds), *The United States and the Multilateral Institutions* (London: Unwin–Hyman, 1990).

2. Paul Taylor, 'Reforming the System: Getting the Money to Talk', in P. Taylor and A. J. Groom, (eds), *International Institutions at Work* (Pinter, 1988), pp. 226–34.

3. *Washington Weekly Report*, XVIII–27, 11 September 1992.

4. World Resources Institute, *World Resources 1992–93*, op. cit., pp. 236–7.

5. UN, *Agenda 21, The United Nations Programme of Action from Rio*, op. cit., Chapter 38. 9, p. 275.

6. This summary is made in a Canadian 'non-paper' titled *Ideas of Some Delegations on Institutional Arrangements for Consideration by the UNCED Preparatory Committee*, Canadian Delegation, Geneva, August 1991, mimeo.

7. Ernst von Weizsacker and Jochen Jesinghaus, *Ecological Tax Reform* (Zed Books, 1992), p. 23.

8. *Washington Weekly Report*, XVII-39, 6 December 1991.

9. See *United Nations News Summary*, NS/35/92, UNIC, London, 15 October 1992.

10. United Nations ST/ADM/SER.B/386, 2 September 1992.

11. Ibid., p. 3.

12. *Washington Weekly Report*, XVIII, 35, 6 November 1992.

13. *Washington Weekly Report*, XIX-30, 7 October 1993.

14. *Washington Weekly Report*, XIX-31, 15 October 1993.

15. *Washington Weekly Report*, XVIII-32, 16 October 1992.

16. Ibid.

17. *Washington Weekly Report*, XVIII-21, 1 July 1992.

18. *Agenda 21*, op. cit., Chapter 33.13, p. 250.

19. Christian Aid, *Aid Report No. 7, British Overseas Aid 1975–1990*, ed. Jessica Woodroffe (Policy Unit, 1991).

20. See *The Independent*, 6 October 1992.

21. *The Independent*, 6 October 1992.

22. M. Grubb *et al.*, op. cit., 1993, p. 175.

23. Ibid., p. 177.

24. Ibid., p. 177.

25. See, 'The Global Environmental Facility', *Our Planet*, Vol. 3, 1991, UNEP, pp. 10–13.

26. Ibid., p. 12.

27. *The Herald* (Glasgow), 9 December 1992.

28. W. Beckerman, *Pricing for Pollution*, 2nd edn, Hobart Paper No. 66, Institute of Economic Affairs, 1990, pp. 13–17.

29. Ibid., p. 29.

30. For different bases of calculations, and comparisons between Japanese, EC and US carbon emissions, expressed in per capita and per unit of GNP terms, see M. Grubb, 'The Greenhouse Effect: Negotiating Targets', *International Affairs*, 66 (1990), pp. 73–4.

31. Ibid., pp. 80–1.

32. A famous example of obfuscation in Scotland during the late 1980s centred on Braemar in Deeside in which, despite the town experiencing several days of temperatures recorded at *circa* 24 F, its pensioner residents were disqualified from making claims for cold-weather payments because temperature readings used as the basis for calculation were taken at Aberdeen, 50 miles east, on the coast.

33. M. Grubb, op. cit., 1990, p. 88.

34. On the apparent irrelevance of natural resource endowments to national prosperity, see J. K. Galbraith, *The Nature of Mass Poverty*, (Harmondsworth: Pelican, 1980), pp. 13–16.

35. See *Our Planet*, Vol. 4, 3, (1992), p. 7.

36. M. Grubb, op. cit., 1990, pp. 73–4.

37. M. Paterson and M. Grubb, 'The international politics of climate change', *International Affairs*, 68 (1992), p. 298.

38. R. Douthwaite, *The Growth Illusion* (Green Books, 1992), pp. 211–13.

39. Grubb, op. cit., 1990, pp. 83–4. For a comprehensive analysis of the question, see also Alan S. Manne and Richard G. Richels, *Buying Greenhouse Insurance; The Economic Costs of Carbon Dioxide Emission Limits* (MIT Press, 1992), especially Chapter 5.

40. Grubb, op. cit., 1990, p. 81.

41. On verification of a climate-change agreement, see Owen Greene, 'Building a global warming convention: lessons from the arms control experience?, in *Pledge and Review Processes: Possible Components of a Climate Convention*, Workshop Report, M. Grubb and N. Steen, RIIA, 1991, pp. xxi–xxxiii. On the IAEA mandate to inspect, see M. F. Imber, *The USA, ILO, UNESCO and IAEA* (London: Macmillan, 1989), Chapter 5.

42. *Agenda 21*, op. cit., Chapter 2. 22, p. 22.

43. United Nations, *Report on the Work of the Organization from the Forty-sixth to the Forty-seventh Session of the General Assembly*, UN, September 1992, pp. 18–19.

44. See *Earth Audit; The World Environment 1972–1992, Where Now?* (UNEP, 1992), pp. 22–3.

45. See A. Rogers, *The Earth Summit* (Los Angeles, Global View, 1993), pp. 238–9.

46. Owen Greene, 'Tackling Global Warming', in P. Smith and K. Warr, (eds), *Global Environmental Issues* (London: Hodder and Stoughton, 1991), p. 192.

47. *The Herald*, Glasgow, 5 July 1993.

48. United Nations A/CONF. 151/PC/64, p. 6

49. Ibid., p. 7.

50. P. Vallely, *Bad Samaritans* (London: Hodder and Stoughton, 1990), p. 301.

51. US General Accounting Office, *Developing Country Debt; Debt Swops for Development and Nature Provide Little Debt Relief*, United States General Accounting Office, GAO/NSAID–92–14, December 1991, p. 9.

52. Ibid., p. 11.
53. Ibid., p. 13.
54. These and many other cases are well summarised by Caroline Thomas, *The Environment in International Relations*, RIIA, 1992, pp. 124–35. See M. F. Imber, 'Environmental security; a task for the UN system', *Review of International Studies* 17, (1991), pp. 201–12. On freshwater questions, see Malin Falkenmark, 'Fresh water as a factor in strategic policy action', in Arthur Westing (ed), *Global Resources and International Conflict* (Oxford: Oxford University Press, 1986), pp. 85–114.
55. The literature on environmental security is well discussed by Thomas, op. cit., 1992, pp. 115–54.
56. These remarks and comparisons are contained in the report prepared by the Secretary-General at the invitation of the Security Council, which places a discussion of new peacekeeping activities in a wider post-Cold War context. See Boutros Boutros-Ghali, *Agenda for Peace* (New York: United Nations, 1992), pp. 6–7.
57. Adapted from tables in D. Deudney, 'The Case Against Linking Environmental Degradation and National Security', in *Millennium*, 19, 1990, pp. 464–6.
58. J. Womack, D. Jones and D. Roos, *The Machine that Changed the World* (Rawson, 1990).

# Bibliography

Adamson, David, *Defending the Earth* (London: I. B. Taurus, 1990).

Anderson, A., 'The environmental aftermath of the Kuwait Iraq Conflict', *Our Planet*, Vol. 3, 1991.

Angell, David J. R., Comer, Justyn and Wilkinson, Matthew L. N., (eds), *Sustaining Earth* (London: Macmillan, 1990).

Axelrod, Robert, *The Evolution of Cooperation* (London: Penguin, 1990).

Barnet, Richard J., *The Lean Years* (London: Abacus, 1980).

Baylis, John and Rengger, N. J. (eds), *Dilemmas of World Politics* (Oxford: Oxford University Press, 1992).

Beckerman, Wilfred, *Pricing for Pollution*, Hobart Paper No. 66, 2nd edn (London: Institute of Economic Affairs, 1990).

Bookchin, Murray, *Post-Scarcity Anarchism* (London: Wildwood, 1974).

Brandt Commission, *Common Crisis North–South: Cooperation for world recovery* (London: Pan, 1983).

Brown, Neville, 'Ecology and World Security', *World Today*, March 1992, pp. 51–4.

Commoner, Barry, *Making Peace with the Planet* (London: Gollancz, 1990).

Denman, D. R., *Markets under the Sea* (London: Institute of Economic Affairs, 1984).

Deudney, Daniel, 'The case against linking environmental degradation and national security', *Millennium*, Vol. 19, no. 3, 1990. pp. 461–76.

Dodson, Andrew, *Green Political Thought* (London: Unwin Hyman, 1990).

Dolman, Antony J. (Ed.), *Global Planning and Resource Management* (Oxford: Pergamon, 1980).

Doran, Peter, 'The Earth Summit (UNCED): Ecology as Spectacle', *Paradigms*, Vol. 7, 1993, pp. 55–66.

Douthwaite, R., *The Growth Illusion* (Green Books, 1992).

Dyer, Hugh 'Environmental Security as a Universal Value: Implications for International Theory' (Leeds, Institute for International Studies, 1992).

*The Ecologist, Blueprint for Survival* (London: Penguin, 1972).

Galbraith, John K., *The Nature of Mass Poverty* (London: Penguin, 1980).

Galbraith, John K., *The Culture of Contentment* (London: Sinclair-Stevenson, 1992).

Galtung, Johan, 'A structural theory of Imperialism', *Journal of Peace Research*, Vol. 13, 1971, pp. 81–94.

Gleik, Peter H., 'Water and conflict; fresh water resources and international security', *International Security*, Vol. 18, no. 1, 1993, pp. 79–112.

George, Susan, *The Debt Boomerang* (Boulder: Westview, 1992).

Gordon, John and Fraser, Caroline, *Institutions and Sustainable Development: Meeting the Challenge* (London: Global Environmental Research Centre, 1991).

Gore, Al, *Earth in the Balance* (London: Earthscan, 1992).

Grubb, Michael, 'The Greenhouse Effect: Negotiating Targets', *International Affairs*, Vol. 66, 1990, pp. 67–89.

Grubb, Michael, Koch, Matthias, Munson, Abby, Sullivan, Francis and Thompson, Koy, *The Earth Summit Agreements* (London: Royal Institute of International Affairs, 1993).

Harrod, J. and Schrijver, Nico (eds), *The UN Under Attack* (Aldershot: Gower, 1988).

Hassan, Shaukat, 'Environmental Issues and Security in South Asia', Adelphi Paper No. 262 (London: International Institute for Strategic Studies, 1991).

Herz, John, *International Politics in the Atomic Age* (Columbia, 1959).

Hewitt, A. and Wells, B. (eds), *Growing out of Debt* (London: Overseas Development Institute, 1989).

Hirschman, A., *Exit, Voice and Loyalty* (Harvard, 1970).

Hollick, Anne, *US Foreign Policy and the Law of the Sea* (Princeton: Princeton University Press, 1981).

Hossain, K. and Chowdhury, S. Roy (eds), *Permanent Sovereignty Over Natural Resources in International Law* (London: Pinter, 1984).

Hurrel, Andrew and Kingsbury, Benedict (eds), *The International Politics of the Environment* (Oxford: Clarendon, 1992).

Illich, Ivan, *Energy and Equity* (London: Calder and Boyars, 1974).

Imber, Mark, *The USA, ILO, UNESCO and IAEA* (London, Macmillan, 1989).

Imber, Mark, 'Environmental Security: A task for the UN system', *Review of International Studies*, Vol. 17, no. 2, April 1991, pp. 201–12.

Imber, Mark, 'Too many cooks? The post-Rio reform of the UN', *International Affairs*, Vol. 69, no. 1, 1993, pp. 55–70.

Imber, Mark, 'The United Nations' role in sustainable development', *Environmental Politics*, Vol. 2, no. 4, 1993, pp. 123–36.

Jacobson, Harold, K., *Networks of Interdependence* (New York: Knopf, 1979).

Johnston, R. J. *Environmental Problems: Nature, Economy and State* (London: Belhaven, 1989).

Kakonen, Jyrki (ed.), *Perspectives on Environmental Conflict and International Politics* (London: Pinter, 1992).

Karns, Margaret and Mingst, Karen (eds), *The United States and the Multilateral Institutions* (London: Unwin-Hyman, 1990).

Keohane, R., *After Hegemony* (Princeton: Princeton University Press, 1984).

Keohane, R and Nye, J., *Power and Interdependence* (New York: Little, Brown, 1977).

Krasner, S. (ed.), *International Regimes* (Ithaca; Cornell University Press, 1983).

Larenti, Jeffrey and Lyman, Francesca, *One Earth, Many Nations* (New York: United Nations Association of the United States of America, 1990).

Leggett, Jonathan, *Global Warming: The Greenpeace Report* (Oxford: Oxford University Press, 1990).

Loescher, Gil, 'Refugee Movements and International Security, Adelphi Paper No. 268, International Institute of Strategic Studies, 1991.

Lowi, Miriam R., 'Bridging the divide: transboundary resource disputes and the case of the West Bank water', *International Security*, Vol. 18, no. 1, 1993, pp. 113–38.

Mann, Alan S. and Richels, Richard G., *Buying Greenhouse Insurance: The Economic Costs of Carbon Dioxide Emission Limits* (Boston, MIT, 1992).

Marples, D. R., *Chernobyl and Nuclear Power in the USSR* (London: Macmillan, 1987).

*Bibliography*

Matthews, Jessica Tuchman, 'Redefining Security', *Foreign Affairs*, Vol. 68, 1989, pp. 162–77.

May, John, *The Greenpeace Book of the Nuclear Age* (London: Gollancz, 1989).

McCleary, Rachel, 'The international community's claim to rights in Brazilian Amazonia', *Political Studies*, Vol. XXXIX, no. 4, 1991, pp. 691–707.

McGrew, A. and Lewis, P. (eds), *Global Politics* (Polity Press, 1990).

Middleton, Neil, O'Keefe, Phil and Moyo, Sam, *Tears of the Crocodile: From Rio to Reality in the Developing World* (London: Pluto, 1993).

*Millennium*, Special Issue: 'Global Environmental Change and International Relations', Vol. 19, no. 3, 1990.

Miller, Morris, *Debt and the Environment: Converging Crises* (New York: United Nations, 1991).

Myers, Norman, *The Gaia Atlas of Future Worlds* (London: Robertson McCarta, 1990).

National Academy of Sciences, *One Earth, One Future* (Washington, DC, National Academy of Sciences, 1990).

Nordhaus, W. D., 'To slow or Not to Slow; the Economics of the Greenhouse Effect', *Economic Journal*, Vol. 101 (1991), pp. 934–6.

Paterson, Matthew, and Grubb, Michael, 'The international politics of climate change', *International Affairs*, Vol. 68, 1992, pp. 293–310.

Pearce, David, Markandya, Anil and Barbier, Edward B., *Blueprint for a Green Economy* (London: Earthscan, 1989).

Pearce, Fred, *Acid Rain* (London: Penguin, 1987).

Poole, J. B. and Guthrie, R. (eds), *Verification Report 1992* (London: VERTIC, annual).

Porter, Gareth and Brown, Janet, *Global Environmental Politics* (Boulder: Westview, 1991).

PRIO, *Environmental Security* (Nairobi: Peace Research Institute; Oslo/UNEP, 1989).

Reisner, Marc, *Cadillac Desert: The American West and its Disappearing Water* (London: Penguin, 1987).

Renninger, J. (ed.), *The Future of the United Nations in an Interdependent World* (Martinus Nijhoff, UNITAR, 1989).

Roberts, Adam, and Kingsbury, Benedict (eds), *United Nations, Divided World* (Oxford: Clarendon, 1993).

Roddick, J., 'UNCED, its "Stakeholders" and the post-UNCED Process' (Glasgow: Institute of Latin American Studies, 1993).

Rogers, Adam, *The Earth Summit* (Los Angeles, Global View, 1993).

Rosenblum, Mort and Williamson, Doug, *Sqandering Eden: Africa at the Edge* (London: Paladin, 1990).

Roxborough, I., *Theories of Underdevelopment* (London: Macmillan, 1979).

Sands, Peter H., *Lessons Learned in Global Governance* (Washington, DC: World Resources Institute, 1990).

Sanger, Clyde, *Ordering the Oceans: The making of the law of the sea* (London: Zed, 1986).

Saetevik, Sunneva, *Environmental Cooperation between the North Sea States* (London: Belhaven, 1988).

Smith, Paul M. and Warr, Kiki (eds), *Global Environmental Issues* (London: Hodder and Stoughton, 1991).

Taylor, Paul, and Groom, A. J. (eds), *International Institutions at Work* (London: Pinter, 1988).

Taylor, Paul and Groom, A. J. R. (eds), *Global Issues in the United Nations* (London: Macmillan, 1989).

Thomas, Caroline, *The Environment in International Relations* (London: Royal Institute of International Affairs, 1992).

Thomas, Caroline (ed.), *Rio, Unravelling the Consequences* (London: Cass, 1994).

United Nations, *Report of the United Nations Conference on the Human Environment* (New York: United Nations, 1973).

United Nations, *The Law of the Sea, Official Text* (New York: United Nations, 1983).

United Nations, *Global Outlook 2000* (New York: United Nations, 1990).

United Nations, *Earth Summit, Agenda 21, The United Nations Programme of Action from Rio* (New York: United Nations, 1993).

United Nations Association of The United States of America, *Washington Weekly Report* (Washington, DC: UNA–USA, weekly).

United Nations Environment Programme, *Our Planet* (Nairobi: UNEP, bi-monthly).

United Nations Environment Programme, *Action for the Environment: The Role of the United Nations* (Nairobi: UNEP, 1989).

United Nations Environment Programme, *Review of the Montevideo Programme for Development and Periodic Review of Environment Law, 1981–1991* (Nairobi: UNEP, 1991).

United Nations Environment Programme, *Annual Report* (Nairobi: UNEP, annual).

United Nations Environment Programme, *State of the Environment* (Nairobi: UNEP, annual).

United States General Accounting Office, *United Nations: US Participation in the Environment Program* (Washington, DC: General Accounting Office, June 1989).

United States General Accounting Office, *United Nations: US Participation in the UN Development Program* (Washington, DC:, General Accounting Office, February 1990).

United States General Accounting Office, *International Environment: International Agreements Are Not Well Monitored* (Washington, DC: General Accounting Office, January 1992).

United States General Accounting Office, *Air Pollution: Difficulties in Implementing a National Air Permit Program* (Washington, DC: General Accounting Office, February 1993).

Vallely, Paul, *Bad Samaritans: First World Ethics and Third World Debt* (London: Hodder and Stoughton, 1990).

Wallerstein, Emmanuel, 'Dependency ... World Systems ... The rise and future demise of the world capitalist system: concepts for comparative analysis', *Comparative Studies in Society and History*, Vol. 16, 1974, pp. 387–415.

*Water Resources Development*, Special Issue: 'Sustainable Water Development', Vol. 4, no. 2, June 1988.

Weale, Albert, *The New Politics of Pollution* (Manchester: Manchester University Press, 1992).

Weir, David, *The Bhopal Syndrome* (London: Earthscan, 1987).

Weizsacker, Ernst von, and Jesinghaus, Jochen, *Ecological Tax Reform* (London: Zed, 1992).

Westing, Arthur H. (ed.), *Herbicides In War* (London: Taylor and Francis, 1984).

Westing, Arthur H. (ed.), *Environment Warfare, A Technical, Legal and Policy Appraisal* (London: Taylor and Francis, 1984).

Westing, Arthur H. (ed.), *Explosive Remnants of War* (London: Taylor and Francis, 1985).

Westing, Arthur H. (ed.), *Global Resources and International Conflict* (Oxford: Oxford University Press, 1986).

Westing, Arthur H. (ed.), *Cultural Norms, War and the Environment* (Oxford: Oxford University Press, 1988).

Westing, Arthur H. (ed.), *Comprehensive Security for the Baltic: An Environmental Approach* (London: Sage, 1989).

Wilkinson, Richard, G., *Poverty and Progress: An ecological model of economic development* (London: Methuen, 1973).

Williams, D., *The Specialized Agencies and the United Nations* (London: Hurst, 1987).

Williamson, John, *The Open Economy and the World Economy* (New York: Basic, 1983).

Womack, James P., Jones, Daniel T., and Roos, Daniel, *The Machine that Changed the World* (New York: Rawson, 1990).

World Commission on Environment and Development, *Our Common Future* (Oxford: Oxford University Press, 1987).

World Resources Institute, *World Resources, A Guide to the Global Environment* (Oxford: Oxford University Press, annual).

Worldwatch Institute, *State of the World* (London: Earthscan, annual).

Young, John, *Post-Environmentalism* (London: Pinter, 1990).

# Index